T0143380

High Efficiency Video Coding and Other Emerging Standards

River Publishers Series in Signal, Image and Speech Processing

Series Editors

MONCEF GABBOUJ
Tampere University of Technology
Finland

THANOS STOURAITIS
University of Patras
Greece

Indexing: All books published in this series are submitted to Thomson Reuters Book Citation Index (BkCI), CrossRef and to Google Scholar

The "River Publishers Series in Signal, Image and Speech Processing" is a series of comprehensive academic and professional books which focus on all aspects of the theory and practice of signal processing. Books published in the series include research monographs, edited volumes, handbooks and textbooks. The books provide professionals, researchers, educators, and advanced students in the field with an invaluable insight into the latest research and developments.

Topics covered in the series include, but are by no means restricted to the following:

- Signal Processing Systems
- Digital Signal Processing
- Image Processing
- Signal Theory
- Stochastic Processes
- Detection and Estimation
- Pattern Recognition
- Optical Signal Processing
- Multi-dimensional Signal Processing
- Communication Signal Processing
- Biomedical Signal Processing
- Acoustic and Vibration Signal Processing
- Data Processing
- Remote Sensing
- Signal Processing Technology
- Speech Processing
- Radar Signal Processing

For a list of other books in this series, visit www.riverpublishers.com

High Efficiency Video Coding and Other Emerging Standards

K. R. Rao

University of Texas at Arlington
USA

J. J. Hwang

Kunsan National University
South Korea

D. N. Kim

Sejong University
South Korea

Published, sold and distributed by:
River Publishers
Alsbjergvej 10
9260 Gistrup
Denmark

River Publishers
Lange Geer 44
2611 PW Delft
The Netherlands

Tel.: +45369953197
www.riverpublishers.com

ISBN: 978-87-93609-03-7 (Hardback)
978-87-93609-02-0 (Ebook)

©2017 River Publishers

All rights reserved. No part of this publication may be reproduced, stored in a retrieval system, or transmitted in any form or by any means, mechanical, photocopying, recording or otherwise, without prior written permission of the publishers.

Foreword

"HIGH EFFICIENCY VIDEO CODING AND OTHER EMERGING STANDARDS", by K. R. Rao, J. J. Hwang and D. N. Kim. 308 pages. River Publishers, 2017. ISBN 978-87-93609-03-7.

Review by Ashraf A. Kassim, Professor, Department of Electrical and Computer Engineering, National University of Singapore, Singapore.

K. R. Rao, Professor of Electrical Engineering at the University of Texas at Arlington and a well-known leading authority in the field of video coding, has teamed up once more with Dr. *J. J. Hwang* of Kunsan National University (South Korea) and Dr. *D. N. Kim* of Barun Technologies Corporation (South Korea) to come up with this new book which focuses on *high efficiency video coding* (HEVC) also known as H.265. This is very relevant today as the proportion of digital video content in on-line Internet data traffic is expected to increase even further in the coming years as major content providers move rapidly into the realm of 4K video. The ability of HEVC to provide huge gains in compression efficiency is expected to speed up its adoption and help it to overtake H.264/AVC which is the de-facto current standard for most on-line video content.

This book is very timely as it discusses various recent developments in the standards arena for HEVC including range extensions and new profiles as well as a number of recently finalized HEVC codecs. It is well positioned to be a comprehensive and up-to-date resource on HEVC as the authors discuss relevant research and reference a number of resources. This includes standards documents, HEVC range extensions and new profiles (3D, multi view, scalability, screen content coding) open source software, review papers, and keynote speeches, but also provide a long list of *projects* that would help graduate students and others to develop a deeper understanding of the HEVC and emerging image coding standards including JPEG LS, JPEG XR, JPEG-XT and JPEG XS. Also presented in detail is *screen content coding* (SCC) which is an extension of HEVC standard with several new tools, including an intra-picture motion estimation/compensation used in natural video coding, palette coding, adaptive color transform, and adaptive motion vector resolution. The authors also provide a glimpse of future video coding beyond HEVC/H.265 and discuss MPEG-4 internet video coding, AVS2 and related codecs including DAALA, THOR, AV1, VP10 (Google), VC1, real media HD, and DSC.

Once again the authors have come up with a very useful resource for researchers, developers and graduate students in the video coding field, enabling them to keep abreast of latest developments.

Dr. Ashraf A. Kassim is a Professor with the Electrical and Computer Engineering Department of the National University of Singapore. His research interests include video/image processing and compression, computer vision and machine learning.

Contents

Preface

Originally this book was planned to be the revised/updated version of our book

"K. R. Rao, D. N. Kim and J. J. Hwang, "Video coding standards: AVS China, H.264/MPEG-4 Part 10, HEVC, VP6, DIRAC and VC-1", published by Springer in 2014. The present book "High efficiency video coding and other emerging standards", stands by itself and is not a II Edition. Chapter 5 is now completely revised and updated. Chapter 8 focusses on screen content coding – HEVC extension. There is no specific reason why this is called Chapter 8.

The main focus now, however, is on High Efficiency Video Coding (HEVC) which is the latest international video coding standard. A detailed description of the tools and techniques that govern the encoder indirectly and the decoder directly is intentionally avoided as there are already number of books (specially by those specialists who are directly involved in proposing/contributing/evaluating/finalizing the detailed processes that constitute the standards) in this field, besides the overview papers, standards documents, reference software, software manuals, test sequences, source codes, tutorials, keynote speakers, panel discussions, reflector and ftp/web sites – all in the public domain. Access to these categories is also provided. Since 2014 in the standards arena in HEVC, range extensions and new profiles (3D, multi view, scalability, screen content coding – SCC) have been finalized. Also others such as MPEG-4 Internet Video Coding (ISO/IEC 14496-33) and AVS2 (IEEE 1857-4) have been standardized. Similarly industry also has finalized codecs such as, DAALA, THOR, VP9 (Google), VC1 (SMPTE), real media HD (Real Networks), and DSC (display stream compression) by VESA (Video Electronics Standards Association). AV1 (Alliance for Open Media, AOM) and VP10 (Google) are being finalized. This book provides access to all these developments.

Brief description of future video coding beyond HEVC/H.265 is provided. About 400 references related to HEVC are added along with 300 projects/problems. The later are self-explanatory and govern the spectrum from a 3-hour graduate credit to research at the masters and doctoral levels. Some require the dedication of groups of researchers with extensive computational (software) and testing facilities. Additional projects/problems based on image coding standards (both developed and some in final stages) such as JPEG LS, JPEG XR, JPEG-XT, JPEG XS (call for proposals for a low-latency lightweight image coding system issued in March 2016 by JPEG) and JPEG-PLENO are added. References, overview papers, panel discussions, tutorials, software, test sequences, conformance bit streams etc. emphasizing these topics are also listed. Brief description related to joint video exploration team (JVET) a.k.a. next generation video coding (NGVC), established by both MPEG and VCEG is targeted for a potential new standard by 2020. Also VESA issued a "call for technology" with the objective to standardize a significantly more complex codec called ADSC (advanced DSC) that is visually lossless at a bit rate lower than DSC. AVS workgroup of China is on a fast forward track in adding SCC capability to AVS2. All these developments can immensely help the researchers, academia and graduate students and provide food for thought to delve deeply into the fascinating world of multimedia compression.

The reader is now well aware that this book is mainly at the research/reference level rather than as a textbook. It challenges the academic/research/industrial community regarding not only the present state-of-the-art but also, more specifically, the future trends and projections. Hence it is an invaluable resource to this community.

Acknowledgements

This book is the result of long-term association of the three authors, K. R. Rao, J. J. Hwang and D. N. Kim. Special thanks go to their respective families for their support, perseverance and understanding. Both Dr. Hwang and Dr. Kim were visiting professors in the multimedia processing lab (MPL) at the University of Texas at Arlington (UTA) whose significant and sustained contributions made this book possible. The first author likes to acknowledge the support provided in various forms by Dr. Peter Crouch, Dean COE, Dr. Jean-Pierre Bardet, Former Dean, College of Engineering (COE), Dr. J. W. Bredow, Chair, Department of Electrical Engineering, and colleagues all in UTA. Dr. G. J Sullivan, Microsoft, Dr. Nam Ling, University of Santa Clara, Dr. Ankur Saxena and Dr. Zhan Ma (both from Samsung Research Labs), Dr. Wen Gao, Peking University, Dr. M. Budagavi, Samsung Research America (SRA), Dr. M. T. Sun, University of Washington, Dr. H. Lakshman Fr. D. Grois of Fraunhofer HHI, Dr. T. Borer, BBC, Dr. Deshpande, Sharp Labs, Dr. Bankoski and Dr. D. Mukherjee (both from Google) Dr. Y. Reznik, Brightcove, Dr. H. Kalva, Florida Atlantic University, Dr. E. Izquierdo, Queen Mary University of London, Dr. E. Magli, Dept. of Electronics and T1elecommunications, Politecnico di Torino, Italy, Dr. W.-K. Cham, Chinese University of Hong Kong, Hong Kong and P. Topiwala, FastVDO for providing various resources in this regard. Constructive review by Dr. Ashraf Kassim, National University of Singapore, is highly valuable. Shiba Kuanar, Harsha Nagathihalli Jagadish and Swaroop Krishna Rao at UTA contributed countless hours in tying up all loose ends (references/copy right releases, proof reading and million other details). The graduate students and alumnae in multimedia processing lab (MPL) at UTA in various ways have made constructive comments.

List of Figures

List of Tables

List of Abbreviations

2D	Two dimension
3D	Three dimension
AAC	Advanced Audio Coding
ACM MoVid	Association for Computer Machinery Mobile Video
ACQP	Adaptive Chroma Quantization Parameter
ACT	Adaptive Color Transform
ADSC	Advanced DSC
AI	All Intra
AHG	Ad Hoc Groups
AIF	Adaptive Interpolation Filter
ALF	Adaptive Loop Filter
AMVP	Advanced Motion Vector Prediction
AOM	Alliance for Open Media
APIF	Adaptive Pre-Interpolation Filter
APSIPA	Asia Pacific Signal and Information Processing Association
AR	Augmented Reality
ARM	Advanced RISC Machines
ASIC	Application-Specific Integrated Circuit
ASIP	Application Specific Instruction Set Processor
ASO	Arbitrary Slice Order
ATR	Average Time Reduction
ATSC	Advanced Television Systems Committee
AVC	Advanced Video Coding—the official MPEG name is ISO/IECIEC 14496-10–MPEG-4 Part 10, and ITU-T name is ITU-T H.264
AVS	Audio Video Standard
AU	Access Unit
BBC	British Broadcasting Corporation
BD	Bjontegaard Distortion
BH	Beyond HEVC
BL	Base Layer
BLA	Broken Link Access
BMSB	Broadband Multimedia Systems and Broadcasting
bpp	Bits per pixel
BRISQUE	Blind/reference-less image spatial quality evaluator

BS	Boundary Strength
BSTM	Butterfly Style Transform Matrices
BTC	Block Truncation Coding
BV	Block vector
BVP	Block vector prediction
CABAC	Context Adaptive Binary Arithmetic Coding
CAVLC	Context-adaptive variable-length coding
CBF	Coded block flag
CCP	Cross Component prediction
CE	Consumer Electronics, Core Experiment
CfE	Call for Evidence
CfP	Call for Proposal
CI	Confidence Interval
CIC	Compound Image Compression
CIF	Common Intermediate Format
COFDM	Co-orthogonal Frequency Division Multiplexing
CPU	Central Processing Unit
CRA	Clean Random Access
CRI	Color Remapping Information
CSVT	Circuits and Systems for Video Technology
CTC	Common Test Conditions
CU	Coding Unit
CUDA	Compute Unified Device Architecture
CWSSIM	Complex-Wavelet Structural Similarity Index
DASH	Dynamic Adaptive Streaming over HTTP
DATE	Design, automation and test in Europe
DCC	Data Compression Conference
DCT	Discrete Cosine Transform
DCTIF	Discrete Cosine Transform Interpolation Filters
DDCT	Directional Discrete Cosine Transform
DF	Deblocking Filter
DIP	Digital Image Processing
DiOR	Digital Operating Room
DIQA	Document Image Quality Assessment
DMB	Digital Multimedia Broadcasting
DMVD	Decoder side Motion Vector Derivation
DMOS	Difference Mean Opinion Score
DPCM	Differential Pulse Code Modulation
DR	DIRAC (BBC)
DSC	Display Stream Compression
DSCQS	Double Stimulus Continuous Quality Scale

DSIS	Double Stimulus Impairment Scale
DSP	Digital Signal Processing
DST	Discrete Sine Transform
DTM	Directional Template Matching
DTT	Discrete Tchebichef Transform
DTV	Digital Television
DVB-H	Digital Video Broadcasting - Handheld
EBU	European Broadcasting Union
EC	Error Concealment
EE	Electrical Engineering
EGK	Exp. Golomb Kth order
EI	Electronic Imaging
EL	Enhancement Layer
EPFL	Ecole Polytechnique Fédérale de Lausanne
ETRI	Electronics and Telecommunications Research Institute
EURASIP	European Association for Signal Processing
FCC	False Contour Candidate
FDAM	Final Draft Amendment
FDIS	Final Draft International Standard
FF	File Format
FIR	Finite Impulse Response
FMO	Flexible Macroblock Ordering
FPGA	Field Programmable Gate Array
fps	Frames per second
FSIM	Feature Similarity
GPU	Graphics Processing Unit
HD	High Definition
HDR	High Dynamic Range
HDTV	High Definition Television
HE-AAC	High Efficiency Advanced Audio Coder
HEIF	High Efficiency Image File Format
HEVC	High efficiency video coding—the official MPEG name is ISO/IEC 23008-2 MPEG-H Part 2 and ITU-T name is ITU-T H.265
HEVStream	High Efficiency Video Stream
HHI	Heinrich Hertz Institute
HLS	High Level Syntax
HM	HEVC Test Model
HOR	Horizontal
HP	High Profile
HTTP	Hyper Text Transfer Protocol
IASTED	International Association of Science and Technology for Development

IBC	Intra Block Copy
ICASSP	International Conference on Acoustics, Speech, and Signal Processing
ICCE	International Conference on Consumer Electronics
ICIEA	IEEE Conference on Industrial Electronics and Applications
ICIP	International Conference on Image Processing
ICME	International Conference on Multimedia and Expo
ICPC	International Conference on Pervasive Computing
ICPR	International Conference on Pattern Recognition
ICT	Integer Cosine Transform
IDR	Intra Decoding Refresh
IEC	International Electrotechnical Commission
IEEE	Institute of Electrical and Electronics Engineers
ILR	Inter Layer Interference
INTDCT	Integer Discrete Cosine Transform
intra HE	Intra High Efficiency
IPTV	Internet Protocol Television
IQA	Image Quality Assessment
IS & T	Information Systems and Technology
ISCAS	International Symposium on Circuits and Systems
ISCCSP	International Symposium on Communications, Control and Signal Processing
ISDB-T	Integrated Services Digital Broadcasting - Terrestrial
ISO	International Organization for Standardization
ISOBMFF	ISO Based Media File Format
ITS	International Telecommunication Symposium
ITU-T	Telecommunication Standardization Sector of the International Telecommunication Union
IVC	Internet Video Coding
ITM	Internet Video Coding Test Model
IVMSP	Image, Video, and Multidimensional Signal Processing
J2K	JPEG 2000
JCI	JND-based compressed image
JCTVC	Joint Collaborative Team on Video Coding
JEM	Joint Exploration Test Model
JETCAS	Journal on Emerging and Selected Topics in Circuits and Systems
JLS	JPEG-LS
JM	Joint Model
JMKTA	JM Key Technology Areas
JND	Just Noticeable Distortion
JPEG	Joint Photographic Experts Group
JPEG-XR	JPEG extended range

JSVM	Joint Scalable Video Model
JTC	Joint Technical Committee
JVCIR	Journal of Visual Communication and Image Representation
JVET	Joint Video Exploration Team
JVT	Joint Video Team
JXR	JPEG-XR
JXT	JPEG-XT
KTA	Key Technology Areas
LAR-LLC	Locally-adaptive Resolution Lossless low-complexity
LCICT	Low Complexity Integer Cosine Transform
LD	Low Delay
LDB	Low Delay with B pictures
LDP	Low Delay with P pictures
LOCO	Low Complexity
LR	Low Resolution
LS	Lossless (near lossless)
LSC	Layer segmentation based coding
L-SEABI	Low complexity back projected interpolation
M.S.	Masters
MANE	Media – Aware Network Element
Mbit/s	Megabits per second
MC	Motion Compensation
MCC	Mixed Content Coding
MCL	Media Communication Lab
MDDCT	Modified Directional Discrete Cosine Transform
MDDT	Mode-Dependent Directional Transform
ME	Motion Estimation
MJPEG	Motion JPEG
MMSP	Multimedia Signal Processing
MOMS	Maximal-Order interpolation with Minimal Support
MOS	Mean Opinion Score
MPEG	Moving Picture Experts Group—the official ISO name is ISO/IEC JTC1/SC29/Working Group 11
Mpixel	Megapixel
Mpm	Most Probable Modes
MRSE	Mean Root Square Error
MSP	Main Still Profile
MV	Motion Vector
MVC	Multi View Coding
NAB	National Association of Broadcasters
NAL	Network Abstraction Layer
NGBT	Next Generation Broadcast Television
NGVC	Next Generation Video Coding

NIQA	Natural image quality assessment
NUH	NAL Unit Header
NTT	Nippon Telegraph and Telephone Corporation
OP	Overview Papers
OR	Outlier ratio
OSS	Open Source Software
P2SM	Pseudo 2D String Matching
PCM	Pulse Code Modulation
PCS	Picture Coding Symposium
PCoIP	PC over IP
PLCC	Pearson linear correlation coefficient
PPS	Picture Parameter Set
PSNR	Peak-to-peak signal to noise ratio
PU	Prediction Unit
PVC	Perceptual Video Coding
PVQ	Perceptual Vector Quantization
QOE	Quality of Experience
QP	Quantizer parameter
RA	Random Access
RADL	Random Access Decodable
RASL	Random Access Skipped
RD	Rate Distortion
R&D	Research and Development
RDOQ	Rate-distortion optimized quantization
RDPCM	Residual Differential Pulse Code Modulation
RDH	Reversible Data Hiding
RICT	Recursive Integer Cosine Transform
RL	Reference Layer
RMHD	Real Media HD
RMVB	Real Media Variable Bitrate
ROI	Region of interest
ROT	Rotational Transform
RTC	Real Time Communications
RTP	Real-time Transport Protocol
RQT	Residual Quad Tree
SAO	Sample adaptive offset
SC	Sub Committee
SC	Stimulus Comparison
SCADA	Supervisory Control and Data Acquisition
SCC	Screen Content Coding
SCM	Screen content coding test module

SDCT	Steerable Discrete Cosine Transform
SDCT-AM	Steerable Discrete Cosine Transform-Alternated Minimization
SDCT-BT	Steerable Discrete Cosine Transform-Binary Tree
SDM	Structural Degradation Model
SDR	Standard Dynamic Range
SDSCE	Simultaneous Double Stimulus Continuous Evaluation
SE	Subjective Evaluation
SEI	Supplemental Enhancement Information
SELC	Sample based weighted prediction for Enhancement Layer Coding
SG	Study Group
SHV	Super Hi-Vision
SHVC	Scalable High Efficiency Video Coding
SI	Switching I
SIQA	Screen image quality assessment
SIQM	Structure Induced Quality Metric
SIMD	Single Instruction Multiple Data
SIP	Signal and Image Processing
SMF	Sparse Matrix Factors
SMPTE	Society of Motion Picture and Television Engineers
SoC	System on Chip
SHV	Super Hi-Vision
SHVC	Scalable HEVC
SP	Switching P
SPA	Signal Processing: Algorithms, Architectures, Arrangements, and Applications
SPEC	Shape primitive extraction and coding
SPL	Special
SPS	Sequence parameter set
Spl H	Special Issues on HEVC
SPIE	Society of Photo-Optical and Instrumentation Engineers
SR	Super Resolution
SROCC	Spearman Rank Order Correlation
SS	Single Stimulus
SSCQE	Single Stimulus Continuous Quality Evaluation
SSIM	Structural Similarity
SSST	Southeastern Symposium on System Theory
SSVC	Spatially Scalable Video Coding
SVC	Scalable Video Coding
SVQA	Subjective VQA
SWP	Sample-based weighted prediction

TB	Transform Block
TE	Tool Experiment
TS	Test Sequence, Transport Stream
TSM	Transform skip mode
TENTM	Tandberg, Ericsson and Nokia Test Model
TMuC	Test Model under Consideration
TMVP	Temporal Motion Vector Prediction
TMO	Tone Mapping Operator
TSA	Temporal Sub-layer Access
TSF	Transform skip flag
TU	Transform Unit
Tut	Tutorial
TX	Texas
TZSearch	Test Zone Search
UHD	Ultra High Definition
UHDTV	Ultra High Definition Television
UTA	University of Texas at Arlington
VC	Video Coding
VCB	Video Coding for Browsers
VCEG	Visual Coding Experts Group-The official ITU name is ITU-T/ SG 16/Q.6- t
VCIP	Visual Communications and Image Processing
VCIR	Visual Communication and Image Representation
VDI	Virtual desktop infrastructure
VDP	Visible difference predictor
VER	Vertical
VESA	Video Electronics Standards Association
ViMSSIM	Video modified Structural Similarity
VLSI	Very Large Scale Integrated circuit
VPS	Video Parameter Set
VQ	Vector Quantization
VQA	Video Quality Assessment
VQEG	Video Quality Experts Group
VSB	Vestigal Sideband
VUI	Video Usability Information
WCG	Wide Color Gamut
WPP	Wavefront Parallel Processing
WD	Working Draft
WG	Working group

WHP	Wavefront based High Parallel
WQVGA	Wide Quarter Video Graphics Array
WSIs	Whole slide images
WVC	Web Video Coding
WVGA	Wide Video Graphics Array
YC_bC_r	Y is the Brightness (luma), C_b is blue minus luma (B-Y) and C_r is red minus luma (R-Y)

5

High Efficiency Video Coding (HEVC)

Abstract

High efficiency video coding, the latest international video-coding standard, is presented. Its comparison with H.264/AVC is cited. Private and national video-coding standards such as DAALA, THOR, DIRAC (BBC), VC1 (SMPTE), VP10 (Google), AV1 (AOM), AVS China (IEEE 1857.4), Real-Media HD (Real Networks), and DSC by VESA are briefly described. Additionally, image-coding standards such as JPEG, JPEG-LS (JLS), JPEG2000, JPEGXR, JPEGXT, JPEG XS, and JPEGPLENO are also included. The focus is on overview of these standards rather than detailed description of the tools and techniques that govern them. A plethora of projects/problems listed at the end of each of these topics challenges the implementation and further research related to them.

Keywords

HEVC, JCT-VC, CU, CTU, PU, TU, SAO, DBF, HM Software, Lossless coding, THOR, DIRAC, VP10, AV1, VC1, AV1, AVS China, JPEG, JPEG-LS, JPEG2000, JPEGXR, JPEGXT, JPEGPLENO, JPEG XS, Beyond HEVC, Transcoders, JVET Subjective evaluation, Legacy codec, Real Media HD, Subjective evaluation, Arithmetic coding, MPEG-DASH, ATSC, SMPTE, unified intra prediction, coding tree unit, prediction unit, transform unit, SAO, coefficient scanning, HM software, lossless coding.

5.1 Introduction

This chapter details the development of HEVC by the JCT-VC.

5.2 Joint Collaborative Team on Video Coding

The JCT-VC is a group of video coding experts from ITU-T Study Group 16 (VCEG) and ISO/IEC JTC 1/SC 29/WG 11 (MPEG) created to develop a new generation video-coding standard that will further reduce by 50% the data rate needed for high-quality video coding, as compared to the current state-of-the-art AVC standard (ITU-T Rec. H.264 | ISO/IEC 14496-10). This new coding standardization initiative is being referred to as HEVC. In ISO/IEC it is called MPEG-H Part2. VCEG is video-coding experts group and MPEG is moving picture experts group. Tan et al. [SE4] have presented the subjective and objective results of a verification test in which HEVC is compared with H.264/AVC. The conclusions in [SE4] are repeated here: "These tests show that the bit rate savings of 59% on average can be achieved by HEVC for the same subjective quality compared with H.264/AVC. The PSNR-based BD-rate average over the same sequences was calculated to be 44%. This confirms that the subjective quality improvements of HEVC are typically greater than the objective quality improvements measured by the method that was primarily used during the standardization process of HEVC. It can therefore be concluded that the HEVC standard is able to deliver the same subjective quality as AVC, while on average (and in the vast majority of typical sequences) requiring only half or even less than half of the bit rate used by AVC."

ITU-T Rec. H.264 | ISO/IEC 14496-10, commonly referred to as H.264/MPEG-4-AVC, H.264/AVC, or MPEG-4 Part 10 AVC has been developed as a joint activity within the JVT. The evolution of the various video coding standards is shown in Figure 5.1.

Footnote: H.265 and recent developments in video coding standards (Seminar presented by Dr. Madhukar Budagavi on 21 Nov. 2014 in the Dept. of Electrical Engineering, Univ. of Texas at Arlington, Arlington, Texas).

Abstract: Video traffic is dominating both the wireless and wireline networks. Globally, IP video is expected to be 79% of all IP traffic in 2018, up from 66% in 2013. On wireless networks, video is 70% of global mobile data traffic in 2013 (Cisco VNI forecast). Movie studios, broadcasters, streaming video providers, TV and consumer electronics device manufacturers are working towards providing immersive "real life" "being there" video experience to consumers by using features such as increased resolution (Ultra HD 4K/8K), higher frame rate, higher dynamic range (HDR), wider color gamut (WCG), and 360 degrees video. These new features along with the explosive growth in video traffic are driving the need for increased compression. This talk will cover basics of video compression and then give an overview of the recently standardized HEVC video-coding standard that provides around 50% higher compression than the current state of the art H.264/AVC video-coding standard. It will also highlight recent developments in the video-coding standards body related to HEVC extensions, HDR/WCG, and discussions on post-HEVC next-generation video coding.

Cisco VNI Global IP Traffic Forecast 2014–2019 predicts that

1) Globally IP video traffic will be 80% of all IP traffic (both business and consumer) by 2019;
2) By 2019 more than 30% of connected flat-panel TV sets will be 4K; and
3) Traffic from wireless and mobile devices will exceed traffic from wired devices by 2016. HEVC is designed to address the increased video traffic and increased resolution such as the 4K and 8K videos.

Latest forecast by Cisco: Video will be 78% of mobile traffic by 2021.

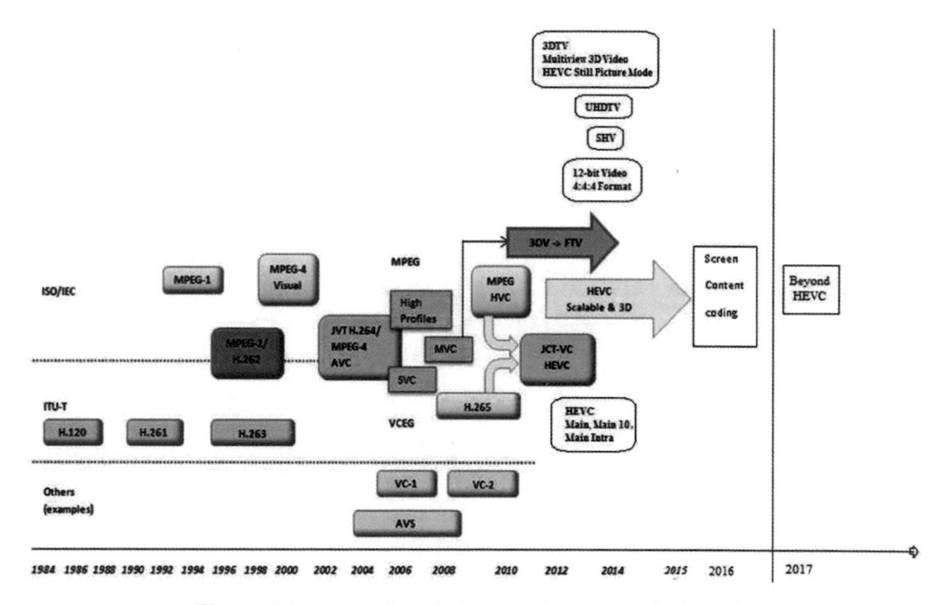

Figure 5.1 Evolution of video coding standards (AOM).

Figure 5.1 Video-coding standardization (courtesy Dr. Nam Ling, Sanfilippo family chair professor, Dept. of Computer Engineering, Santa Clara University, Santa Clara, CA, USA) [E23], AOM.

The JCT-VC is co-chaired by Jens-Rainer Ohm and Gary Sullivan, whose contact information is provided below.

ITU-T Contact for JCT-VC	**Meetings**
Mr Gary SULLIVAN Rapporteur, Visual coding Question 6, ITU-T Study Group 16 Tel: +1 425 703 5308 Fax: +1 425 936 7329 E-mail: garysull@microsoft.com Mr Thomas WIEGAND Associate Rapporteur, Visual coding Question 6, ITU-T Study Group 16 Tel: +49 30 31002 617 Fax: +49 30 392 7200 E-mail: thomas.wiegand@microsoft.com	Future meetings Geneva, Switzerland, October 2013 (tentative) Vienna, Austria, 27 July–2 August 2013 (tentative) Incheon, Korea, 20–26 April 2013 (tentative) Geneva, Switzerland, 14–23 January 2013 (tentative)
ISO/IEC contacts for JCT-VC	
Mr Jens-Rainer OHM Rapporteur, Visual coding Question 6, ITU-T Study Group 16 Tel: +49 241 80 27671 E-mail: ohm@ient.rwth-aachen.de	Mr Gary SULLIVAN Rapporteur, Visual coding Question 6, ITU-T Study Group 16 Tel: +1 425 703 5308 Fax: +1 425 936 7329 E-mail: garysull@microsoft.com

Additional information can be obtained from

http://www.itu.int/en/ITU-T/studygroups/com16/video/Pages/jctvc.aspx
JCT-VC has issued a joint call for proposals in 2010 [E7]

- 27 complete proposals submitted (some multi-organizational).
- Each proposal was a major package—lots of encoded video, extensive documentation, extensive performance metric submissions, sometimes software, etc.
- Extensive subjective testing (3 test labs, 4,200 video clips evaluated, 850 human subjects, and 300,000 scores)
- Quality of proposal video was compared to AVC (ITU-T Rec. H.264 | ISO/IEC 14496-10) anchor encodings
- Test report issued **JCTVC-A204/N11775**
- In a number of cases, comparable quality at half the bit rate of AVC (H.264)

- Source video sequences grouped into five classes of video resolution from quarter WVGA (416 × 240) to size 2,560 × 1,600 cropped from 4k × 2k ultra HD (UHD) in YC_bC_r 4:2:0 format progressively scanned with 8 bpp.
- Testing for both "RA" and "low delay" (no picture reordering) conditions.

Table 5.1 Test Classes and Bit Rates (constraints) used in the CfP [E7]

Class	Bit Rate 1	Bit Rate 2	Bit Rate 3	Bit Rate 4	Bit Rate 5
A: 2560 × 1600p30	2.5 Mbit/s	3.5 Mbit/s	5 Mbit/s	8 Mbit/s	14 Mbit/s
B1: 1080p24	1 Mbit/s	1.6 Mbit/s	2.5 Mbit/s	4 Mbit/s	6 Mbit/s
B2: 1080P.50-60	2 Mbit/s	3 Mbit/s	4.5 Mbit/s	7 Mbit/s	10 Mbit/s
C: WVGAp30-60	384 kbit/s	512 kbit/s	768 kbit/s	1.2 Mbit/s	2 Mbit/s
D: WQVGAp30-60	256 kbit/s	384 kbit/s	512 kbit/s	850 kbit/s	1.5 Mbit/s
E: 720p60	256 kbit/s	384 kbit/s	512 kbit/s	850 kbit/s	1.5 Mbit/s

Figures 5.2 and 5.3 show results averaged over all of the test sequences; in which the first graph (Figure 5.2) shows the average results for the RA constraint conditions, and the second graph (Figure 5.3) shows the average results for the low delay constraint conditions.

The results were based on an 11-grade scale, where 0 represents the worst and 10 represents the best individual quality measurements. Along with each MOS data point in the figures, a 95% CI is shown.

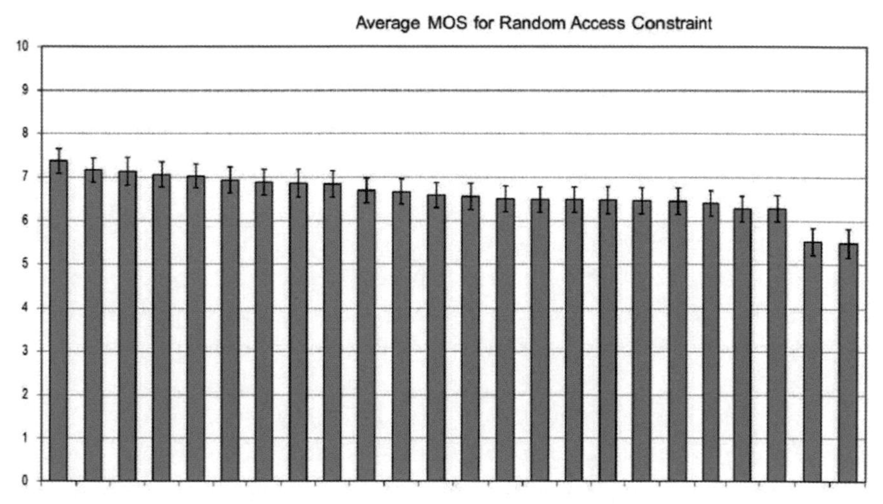

Figure 5.2 Overall average MOS results over all classes for RA-coding conditions [E7].

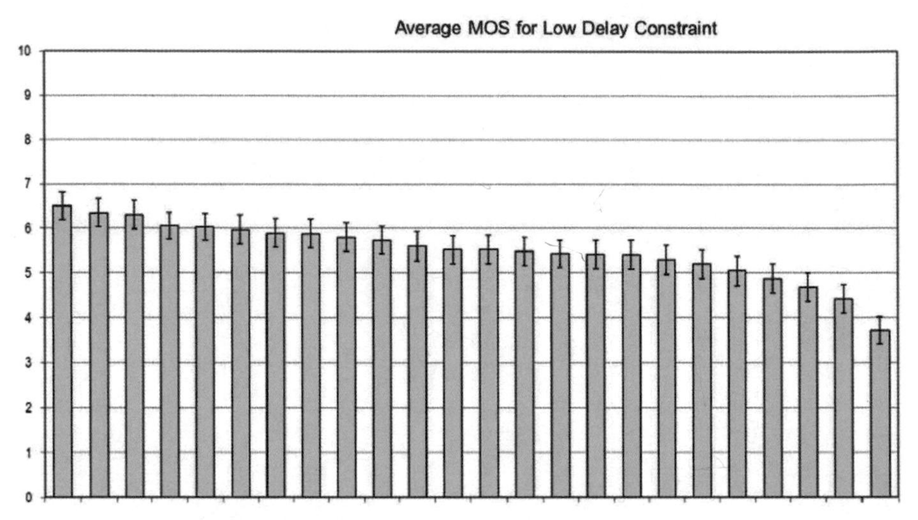

Figure 5.3 Overall average MOS results over all classes for low delay coding conditions [E7].

A more detailed analysis performed after the tests, shows that the best-performing proposals in a significant number of cases showed similar quality as the AVC anchors (H.264/AVC) at roughly half the anchor bit rate [E25, E61, E99]. More recently, Tan et al. [SE4] present the subjective and objective results of a verification test in which the performance of HEVC Main profile is compared with H.264/AVC HP using video test sequences at various bit rates and resolutions. They conclude that the HEVC is able to deliver the same subjective quality as H.264/AVC while on average requiring only half or even less than half the bit rate used by the latter. Tan et al. [SE4] describe in detail, the methodology, test procedures, test sequences, comparison metrics, and other factors in arriving at the conclusions.

The technical assessment of the proposed technology was performed at the first JCT-VC meeting held in Dresden, Germany, 15–23 April 2010. It revealed that all proposed algorithms were based on the traditional hybrid coding approach, combining motion-compensated prediction between video frames with intra-picture prediction, closed loop operation with in-loop filtering, 2D transform of the spatial residual signals, and advanced adaptive entropy coding.

As an initial step toward moving forward into collaborative work, an initial TMuC document was produced, combining identified key elements from a group of seven well performing proposals. This first TMuC became the basis of a first software implementation, which after its development has begun to enable more rigorous assessment of the coding tools that it contains as well as additional tools to be investigated within a process of "TE" as planned at the first JCT-VC meeting.

P.S.: Detailed subjective evaluations in mobile environments (smart phone/ iPad, Tablet), however, have shown that the user (observer) experience is not significantly different when comparing H.264/AVC and HEVC compressed video sequences at low bit rates (200 and 400 kbps) and small screen sizes [E124, E210]. Advantages of HEVC over H.264/AVC appear to increase dramatically at higher bit rates and high resolutions such as HDTV, UHDTV, etc.

One of the most beneficial elements for higher compression performance in high-resolution video comes due to introduction of larger block structures with flexible mechanisms of sub-partitioning. For this, the TMuC defines *coding units* (CUs) which define a sub-partitioning of a picture into rectangular regions of equal or (typically) variable size. The coding unit replaces the macroblock structure (H.264) and contains one or several *prediction unit(s)* (PUs) and *transform units* (TUs). The basic partition geometry of all these elements is encoded by a scheme similar to the quad-tree segmentation structure. At the level of PU, either intra-picture or inter-picture prediction is selected.

The paper "Block partitioning structure in the HEVC standard", by I.-K. Kim et al. [E93], explains the technical details of the block partitioning structure and presents the results of an analysis of coding efficiency and complexity.

- Intra-picture prediction is performed from samples of already decoded adjacent PUs, where the different modes are DC (flat average), horizontal, vertical, or one of up to 28 angular directions (number depending on block size), plane (amplitude surface) prediction, and bilinear prediction. The signaling of the mode is derived from the modes of adjacent PUs.
- Inter-picture prediction is performed from region(s) of already decoded pictures stored in the reference picture. This allows selection among multiple reference pictures, as well as bi-prediction (including weighted averaging) from two reference pictures or two positions in the same reference picture. In terms of the usage of the motion vector (quarter pixel precision), merging of adjacent PUs is possible, and non-rectangular sub-partitions are also possible in this context. For efficient encoding, skip and direct modes similar to the ones of H.264/AVC are defined,

and derivation of motion vectors from those of adjacent PUs is made by various means such as median computation or a new scheme referred to as *motion vector competition*.

At the TU level (which typically would not be larger than the PU), an integer spatial transform similar in concept to the DCT is used, with a selectable block size ranging from 4 × 4 to 64 × 64. For the directional intra modes, which usually exhibit directional structures in the prediction residual, special *mode-dependent directional transforms* (*MDDT*) [E8, E51, E17, E301], [DCT19]] are employed for block sizes 4 × 4 and 8 × 8. Additionally, a *rotational transform (See p.5.13)* can be used for the cases of block sizes larger than 8 × 8. Scaling, quantization and scanning of transform coefficient values are performed in a similar manner as in AVC.

At the CU level, it is possible to switch on an *adaptive loop filter* (*ALF*) which is applied in the prediction loop prior to copying the frame into the reference picture buffer. This is an FIR filter which is designed with the goal to minimize the distortion relative to the original picture (e.g., with a least-squares or Wiener filter optimization). Filter coefficients are encoded at the slice level. In addition, a DF (similar to the DF design in H.264/AVC) [E73] is operated within the prediction loop. The display output of the decoder is written to the decoded picture buffer after applying these two filters. Please note that the ALF has been dropped in the HEVC standard [E235, E61, E99]. In the updated version, in loop filtering consists of deblocking [E73, E209, E283, E357, E378] and SAO filters (Figure 5.4). See [E87, E111, E373] about SAO in the HEVC standard.

The TMuC defines two context-adaptive entropy coding schemes, one for operation in a lower-complexity mode, and one for higher-complexity mode.

A software implementation of the TMuC has been developed. On this basis, the JCT-VC is performing a detailed investigation about the performance of the coding tools contained in the TMuC package, as well as other tools that have been proposed in addition to these. Based on the results of such *TEs*, the group will define a more well-validated design referred to as a TM as the next significant step in HEVC standardization. Specific experiments have been planned relating to a tool-by-tool evaluation of the elements of the current TMuC, as well as evaluation of other tools that could give additional benefit in terms of compression capability or complexity reduction in areas such as intra-frame and inter-frame prediction, transforms, entropy coding and motion vector coding. Various *ad hoc groups* (AHGs) have been set up to perform additional studies on issues such as complexity analysis, as listed below:

***Ad Hoc* Coordination Groups Formed**

- JCT-VC project management
- Test model under consideration editing
- Software development and TMuC software technical evaluation
- Intra-prediction
- Alternative transforms
- MV precision
- In-loop filtering
- Large block structures
- Parallel entropy coding

In summary, the addition of number of tools within the MC transform prediction hybrid-coding framework (adaptive intra/inter-frame coding, adaptive directional intra-prediction, multiple block size motion estimation, SAO filter [E87, E111, E373], in loop DF [E73, E209, E283, E302, E357, E378, E398], entropy coding (CABAC, see [E67, E68]), multiple frame MC-weighted prediction, integer transforms from 4×4 to 32×32 [E8, E74], Hadamard transform coding of dc coefficients in intra frame coding introduced in H.264/AVC and various other tools as enumerated below has shown further gains in the coding efficiency, reduced bit rates, higher PSNR, etc. compared to H.264/AVC. Thus, HEVC holds promise and potential in a number of diverse applications/fields and is expected to eventually overtake H.264/AVC.

a) Directional prediction modes (up to 34) for different PUs in intra prediction (Figure 5.4) [E49, E104].
b) Mode dependent directional transform (MDDT)$^{+}$ besides the traditional HOR/vertical scans for intra frame coding [E17, E51, E301], [DCT15, DCT19, DCT32].
c) Rotational transforms for block sizes larger than (8×8) (see P.5.13).
d) Large size transforms up to 32×32 [E8, E74].
e) In loop-DF [E73, E209, E283, E302, E357, E378, E398, E415] and SAO filter [E87, E111, E373].
f) Large size blocks for ME/MC.
g) Non-rectangular motion partition [E61].

Note (b) and (c) have been dropped as they contribute very little to coding efficiency at the cost of substantial increase in complexity. Other directional transforms that were proposed, like directional DCT [E112] and MDDCT/DST [E113], were also dropped due to the same reason. Recently steerable DCT (SDCT) and its integer approximation INTSDCT that can be steered in any direction are proposed [E392]. The authors state “SDTC allows to rotate in a flexible way pairs of basis vectors and enables precise matching of directionality in each image block, achieving improved coding efficiency." They have implemented SDCT in image coding and compared with DCT and DDCT. They also suggest a possible implementation of SDCT and INTSDCT in a video compression standard such as HEVC. See the projects P.5.262 thru P.5.267. Both in loop deblocking and SAO filters reduce blocking and ringing artifacts. However, in HD and UHD videos, false contour is a major artifact due to unbalanced quantization in smooth areas. Huang et al. [E393] have identified the cause for false contours and have developed effective method for detection and removal of false contours in both compressed images and videos while preserving true edges and textures.

It is interesting to note that some of these tools such as rotational transforms are reconsidered in the NGVC (beyond HEVC) [BH2]. Other tools considered in NGVC (proposed earlier in HEVC) are:

- Large blocks (both CU and TU sizes);
- Fine granularity intra prediction angles (Figure 3);
- Bi-directional optical flow (Figures 4 and 5) [9];
- Secondary transform, both implicit and explicit (Figures 7 and 8) [10, 11];
- Multi-parameter Intra prediction (Figure 9) [12];
- Multi-hypothesis CABAC probability estimation [13].

References and figures cited here are from Alshin et al. [BH2].

P.S.: The introduction to JCT-VC is based on the paper by G.J. Sullivan and J.-R. Ohm published in applications of digital image processing XXXIII, proc. of SPIE vol. 7798, pp. 7798V-1 through 7798V-7, 2010. Paper title is "Recent developments in standardization of high efficiency video coding (HEVC)" [E7].

For the recent developments in HEVC, the reader is referred to an excellent review paper:

G.J. Sullivan et al., "Overview of high efficiency video coding (HEVC) standard", IEEE Trans. CSVT, vol. 22, pp. 1649–1668, Dec. 2012 [E61]. Also keynote speeches on HEVC [E23, E25, E99]. Also tutorials on HEVC (see the section on tutorials). Also see HEVC text specification draft 8 [E60]. An updated paper on HEVC is G.J. Sullivan et al., "Standardized extensions of high efficiency video coding (HEVC)", IEEE Journal of selected topics in signal processing, vol. 7, pp. 1001–1016, Dec. 2013 [E160]. Another valuable resource is the latest book: V. Sze, M. Budagavi and G.J. Sullivan (editors) "High efficiency video coding (HEVC): algorithms and architectures", Springer, 2014 [E202]. Another valuable book is M. Wien,

"High Efficiency Video Coding: Coding Tools and Specification", Springer, 2015. (See Section on books on HEVC).

$^{+}$For intra-mode, an alternative transform derived from DST is applied to 4×4 luma blocks only. For all other cases, integer DCT is applied.

5.3 Analysis of Coding Tools in HEVC Test Model, HM 1.0: Intra-Prediction

In HM 1.0, unified intra prediction provides up to 34 directional prediction modes for different PUs. With the PU size of 4×4, 8×8, 16×16, 32×32 and 64×64, there are 17, 34, 34, 34 and 5 prediction modes available respectively. The prediction directions in the unified intra prediction have the angles of $\pm$ [0, 2, 5, 9, 13, 17, 21, 26, 32]/32. The angle is given by displacement of the bottom row of the PU and the reference row above the PU in case of vertical prediction, or displacement of the rightmost column of the PU and the reference column left from the PU in case of horizontal prediction. Figure 5.4 shows an example of prediction directions for 32×32 block size. Instead of different accuracies for different sizes, the reconstruction of the pixel uses the linear interpolation of the reference top or left samples at 1/32th pixel accuracy for all block sizes.

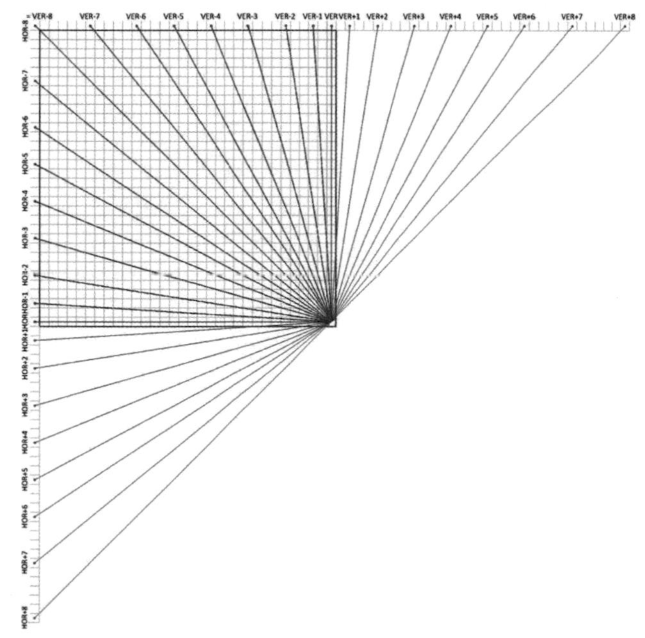

Figure 5.4 Available prediction directions in the unified intra-prediction in HM 1.0.

More details on unified intra prediction in HM 1.0 are available in http://www.h265.net/2010/12/analysis-of-coding-tools-in-hevc-test-model-hm-intra-prediction.html

The working draft (WD) of the HEVC has gone through several updates/revisions and the final draft international standard (FDIS) has come out in January 2013. This refers to Main, Main10 and Main Intra-profiles. In August 2013, five additional profiles Main 12, Main 4:2:2 12, Main 4:4:4 10, and Main 4:4:4 12 were released [E160]. Other range extensions include increased emphasis on high quality coding, lossless coding and screen content coding. See the section on references on screen content coding. Scalable video coding (spatial, temporal, quality, hybrid, and color gamut scalabilities) and multiview video coding were finalized in July 2014 and standardized in October 2014. See section on special issues on HEVC.

IEEE Journal on Emerging and Selected Topics in Circuits and Systems (*JETCAS*) has published the special Issue on Screen Content Video Coding and Applications in December 2016. See the overview paper on SCC [OP11 and SCC48]. This special issue opens up number of research areas in SCC. (See p.5.85 and p.5.85a).

Scalability extensions [E160, E325, E326, E410] and 3D video extensions which enable stereoscopic and multiview representations and consider newer 3D capabilities such as the depth maps and view-synthesis techniques have been finalized in 2015. For work on 3D video topics for multiple standards including 3D video extensions JCT-VC formed a team known as JCT on 3D Video (JCT-3V) in July 2012 [E160]. The overall objective is to reduce the bit rate, increase the PSNR significantly compared to H.264/AVC with reasonable increase in encoder/decoder complexity. These three overview papers [E325, E326, E160] give detailed development and description of these extensions. See also references listed under overview papers.

5.4 HEVC Encoder

IEEE Trans. CSVT vol. 22, Dec. 2012 is a special issue on emerging research and standards in next generation video coding [E45]. This special issue provides the latest developments in HEVC-related technologies, implementations and systems with focus on further research. As the HEVC development is ongoing, this chapter concludes with a number of projects related to HEVC with appropriate references and the info on the KTA and HEVC software [E97]. Hopefully these projects can provide additional insights to the tools and techniques proposed and provide a forum for their modifications leading to further improvements in the HEVC encoder/decoder. Figure 5.5 describes the HEVC encoder block diagram [E61]. This does not show the coded bit stream representing the various modes (intra/inter, CU/PU/TU sizes, intra angular prediction directions/modes, MV prediction, scaling and quantization of transform coefficients and other modes are shown in the decoder block diagram—see Figure 5.6).

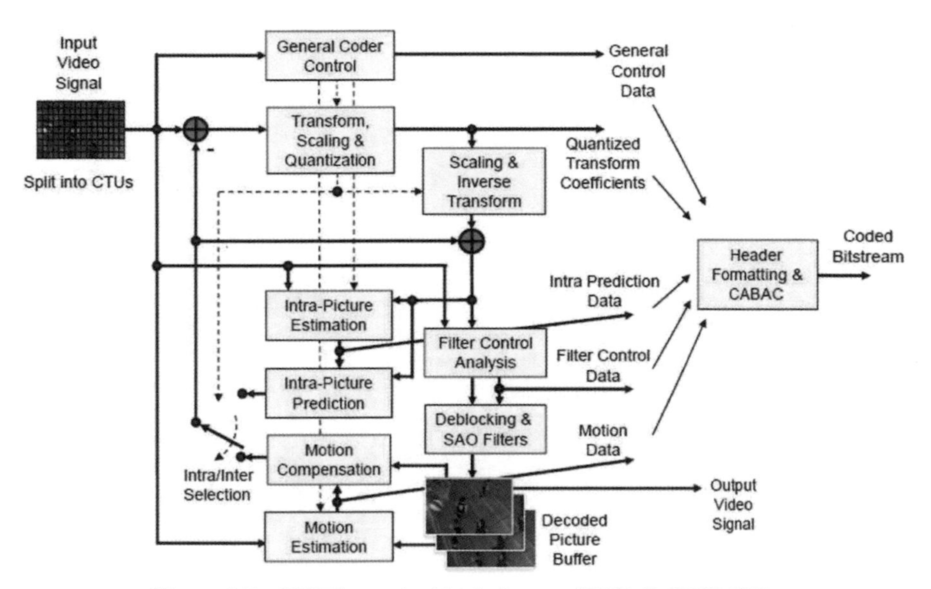

Figure 5.5 HEVC encoder block diagram [E59] © IEEE 2012.

For directional intra modes, an alternative transform related to DST is applied to 4×4 luma prediction residuals. For all other cases integer DCT is applied (Intra/Inter-chroma, and Inter-luma). This change was adopted in Stockholm in July 2012.

In entropy coding only the context adaptive binary arithmetic coding (CABAC) [E67, E68, E390] is adopted unlike two (CAVLC and CABAC) in H.264/AVC. Details on context modeling, adaptive coefficient scanning, and coefficient coding are provided in [E58]. Chapter 8—Entropy coding in HEVC—authored by Sze and Marpe in [E202] describes the functionality and design methodology behind CABAC entropy coding in HEVC. In the conclusions section of this Chapter Sze and Marpe state “The final design of CABAC in HEVC shows that by accounting for implementation cost and coding efficiency when designing entropy coding algorithms results in a design that can maximize processing speed and minimize area cost, while delivering high coding efficiency in the latest video coding standard'. The references cited at the end of this chapter are extensive and valuable.

Mode dependent directional transform [E17, E51, E301] is not adopted. Only INTDCT (separable 2-D) for all cases other than 4 × 4 intra luma is used.

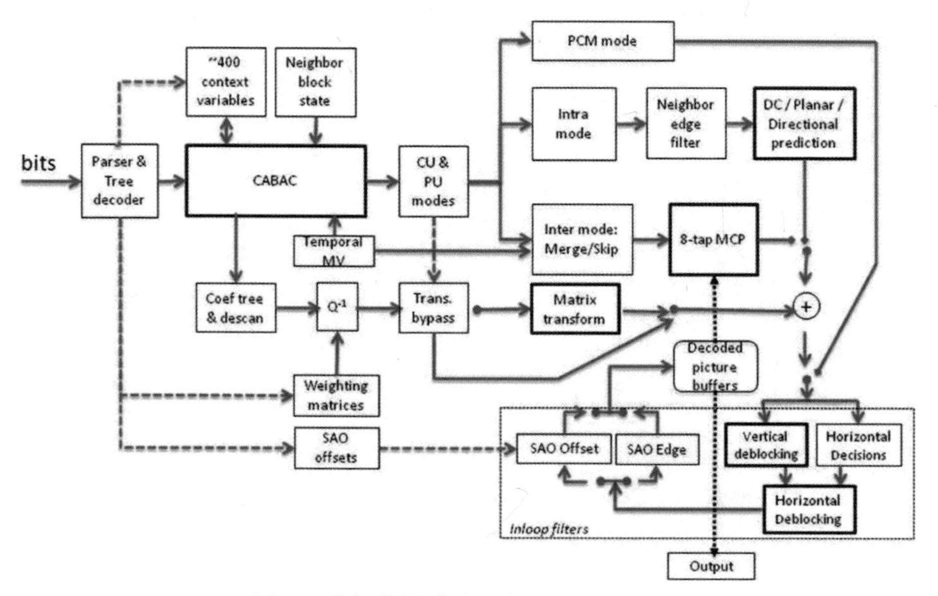

Figure 5.6 HEVC decoder block diagram.

+The decoder block diagram is adopted from C. Fogg, "Suggested figures for the HEVC specification", ITU-T/ISO-IEC Document: JCTVC-J0292r1, July 2012.

While the HEVC follows the traditional (also proven) block based motion compensated prediction followed by transform, quantization, variable length coding (also adaptive intra mode) it differs significantly by adopting flexible quad tree coding block partitioning structure. This recursive tree coding structure is augmented by large block size transforms (up to 32 × 32), advanced motion prediction, sample adaptive offset (SAO) [E87, E111, 373] besides the DF [E73, E209, E283, E302, E357, E378, E398]. The large multiple sizes recursive structure is categorized into coding unit (CU), prediction unit (PU) and transform unit (Figure 5.7a) [E44]. For details about transform coefficient coding in HEVC see [E74].

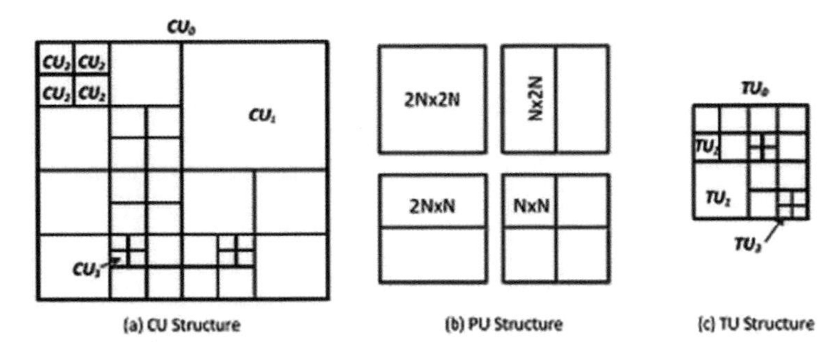

Figure 5.7a Recursive block structure for HEVC, where k indicates the depth for CU_k and TU_k [E44].

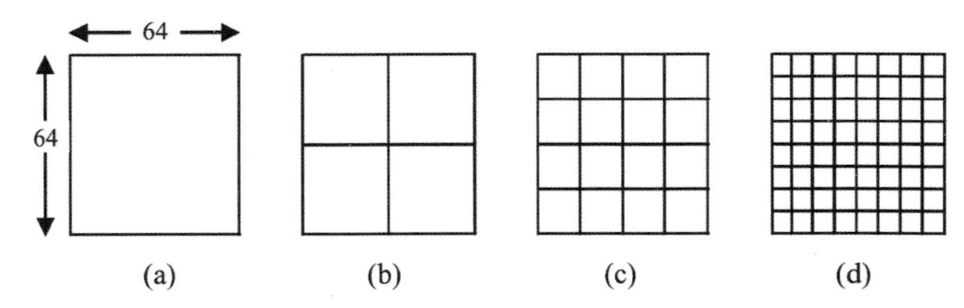

Figure 5.7b Quadtree partitioning of a 64 × 64 CTU. (i) 64 × 64 CU at depth 0. (ii) Four CUs of 32 × 32 at depth 1. (iii) 16 CUs of 16 × 16 at depth 2, and (iv) 64 CUs of 8 × 8 at depth 3.

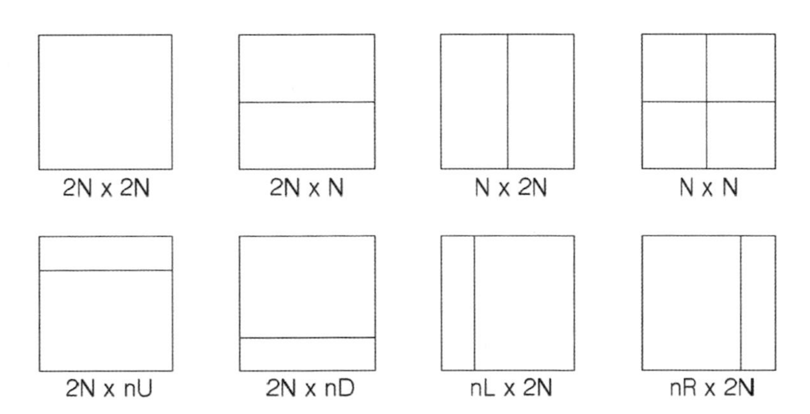

Figure 5.7c Partition mode for the PU of a CU.

In HEVC, the coding tree unit (CTU) is the basic unit of the quadtree coding method, and prediction and transform are performed at each coding unit (CU) that is a leaf node of the tree. The sizes of CTUs can be 16 × 16, 32 × 32, or 64 × 64 pixels, and CU sizes can be 8 × 8, 16 × 16, 32 × 32, or 64 × 64 pixels. Figure 5.7b shows CU formations in a 64 × 64 CTU and each CU can be partitioned into prediction units (PUs), as shown in Figure 5.7c. There are a number of modes, such as skip/merge $2N \times 2N$, inter $2N \times 2N$, inter $N \times N$, inter $N \times 2\ N$, inter $2N \times N$, inter $2N \times nD$, inter $2N \times nU$, inter $nL \times 2N$, inter $nR \times 2N$, intra $2N \times 2N$, and intra $N \times N$. For each CU, the mode having the minimum rate–distortion (RD) cost is called the best mode, which is used as a competitor to decide the quadtree partition structure of the CTU. The decision process of the best mode includes prediction, transform, quantization, and entropy coding, which usually requires high-computational cost.

Complexity of the HEVC encoder is further increased by introducing intra adaptive angular direction prediction, mode dependent context sample smoothing, adaptive motion parameter prediction, in loop filtering (DF and sample adaptive offset -SAO). Details on these two in loop filters and their effectiveness in improving the subjective quality are described in Chapter 7—In-loop filters in HEVC authored by Norkin et al., see [E202]. These and other tools contribute to 50% improved coding efficiency over H.264 at the cost of substantial increase in encoder complexity. The decoder complexity, however, is similar to that of H.264/AVC [E23, E61, E105]. Several techniques for reducing the intra prediction encoder complexity (see [E42] and various references cited at the end) are suggested. See also [E108]. Zhang and Ma [E42, E147] are also exploring the reduction of inter prediction complexity. They suggest that these two areas (intra and inter prediction modes) can be combined to reduce the overall HEVC encoder complexity. (See projects P.5.16 thru P.5.19). This is a fertile ground for research. A summary of the tools included in main and high efficiency 10 [HE10] is shown in Table 5.2. Details of these tools are described in Test Model encoder description [E59]. See also the review papers [E61, E99, E107]. The paper [E61]] states "To assist industry community in learning how to use the standard, the standardization effort not only includes the development of a text specification document (HM8) but also reference software source code (both encoder/decoder)" [E97]. This software can be used as a research tool and as the basis of products. This paper also states "A standard test data suite is also being developed for testing conformance to the standard."

Table 5.2 Structure of tools in HM9 configuration

Main	High efficiency 10 (HE10)
High-level Structure	
High-level support for frame rate temporal nesting and RA	
Clean RA (CRA) support	
Rectangular tile-structured scanning	
Wave front-structured processing dependencies for parallelism	
Slices with spatial granularity equal to coding tree unit	
Slices with independent and dependent slice segments	
Coding units, Prediction units, and Transform units:	
Coding unit quadtree structure square coding unit block sizes $2N \times 2N$, for $N = 4, 8, 16, 32$ (i.e., up to 64×64 luma samples in size)	
Prediction units (for coding unit size $2N \times 2N$: for Inter, $2N \times 2N$, $2N \times N$, $N \times 2N$, and, for $N > 4$, also $2N \times (N/2 + 3N/2)$ and $(N/2 + 3N/2) \times 2N$; for Intra, only $2N \times 2N$ and, for $N = 4$, also $N \times N$)	
Transform unit tree structure within coding unit (maximum of 3 levels)	
Transform block size of 4×4 to 32×32 samples (always square)	
Spatial Signal Transformation and PCM Representation	
DCT-like integer block transform; for Intra also a DST-based integer block transform (only for Luma 4x4)	
Transforms can cross prediction unit boundaries for Inter; not for Intra	
Skipping transform is allowed for 4×4 transform unit	
PCM coding with worst-case bit usage limit	
Intra-picture Prediction:	
Angular intra prediction (35 modes including DC and Planar)	
Planar intra prediction	
Inter-picture Prediction:	
Luma motion compensation interpolation: 1/4 sample precision, 8×8 separable with 6 bit tap values for 1/2 precision, 7×7 separable with 6 bit tap values for 1/4 precision	
Chroma motion compensation interpolation: 1/8 sample precision, 4×4 separable with 6 bit tap values	
Advanced motion vector prediction with motion vector "competition" and "merging"	
Entropy Coding:	
Context adaptive binary arithmetic entropy coding (CABAC)	
Rate-distortion optimized quantization (RDOQ)	
Picture Storage and Output Precision:	
8 bit-per-sample storage and output	10 bit-per-sample storage and output
In-Loop Filtering:	
Deblocking Filter (DF)	
Sample-adaptive offset filter (SAO)	

5.4.1 Intra Prediction

Figure 5.8 shows the 33 intra prediction angle directions [E59, E61, E125] corresponding to the VER and HOR described in Figure 5.4. Figure 5.9 shows the 33 intra prediction mode directions. The mapping between the intra prediction mode directions and angles is shown in Table 5.3 [E125]. See also [E102]. These intra prediction modes contribute significantly to the improved performance of HEVC. Statistical analysis of the usage of the directional prediction modes for AI case has shown that besides planar (mode 0) and dc (mode 1), HOR (mode 10) and vertical (mode26) are at the top of this ranking [E104]. The authors in [E104] by developing a new angular table have demonstrated improved coding gains for video sequences with large amounts of various textures. Each intra-coded PU shall have an intra-prediction mode for luma and another for chroma components. All TUs within a PU shall use the same associated mode for each component. Encoder then selects the best luma intra-prediction mode from the 35 directions (33 plus planar and DC). Due to increased number of directions (compared to those in H.264/AVC – Chapter 4) HEVC considers 3 mpm compared to 1 mpm in H.264/AVC. For chroma of intra-PU, encoder then selects the best chroma prediction mode from 5 modes including planar, DC, HOR, vertical and direct copy of intra prediction mode for luma. Details about mapping between intra prediction direction and mode # for chroma are given in [E59, E125].

Detailed description of the HEVC encoder related to slices/tiles, coding units, prediction units, transform units, inter-prediction modes, special coding modes, MV estimation/prediction, interpolation filters for fractional pixel MV resolution, weighted prediction, transform sizes (4×4, 8×8, 16×16, and 32×32) [E74], scanning the transform coefficients – Figure 5.10 – (see [E74] for details), scaling/quantization, loop filtering (deblocking filter [E73, E209, E283, E357, E378, E398]) and SAO [E87, E111, E373, E398]), entropy coding (CABAC [E67, E68, E390]) and myriad other functionalities are provided in [E61, E99, E107]. An excellent resource is V. Sze, M. Budagavi and G.J. Sullivan (Editors), “High efficiency video coding: Algorithms and architectures”, Springer 2014. [E202]. Another excellent resource is the book M. Wien, “High efficiency video coding: Coding tools and specification”, Springer 2015. In [E202], various aspects of HEVC are dealt within different chapters contributed by various authors who have been involved in all phases of the HEVC development as an ITU-T and ISO/IEC standard. [E202] Popular zigzag scan is not adopted.

P.S.: Hsu and Shen [E357] have developed the VLSI architecture and hardware implementation of a highly efficient DF for HEVC that can achieve 60 fps video (4K×2K) under an operating frequency of 100 MHz. They also list several references related to DF (H.264/AVC and H.265) and SAO filter (H.265).

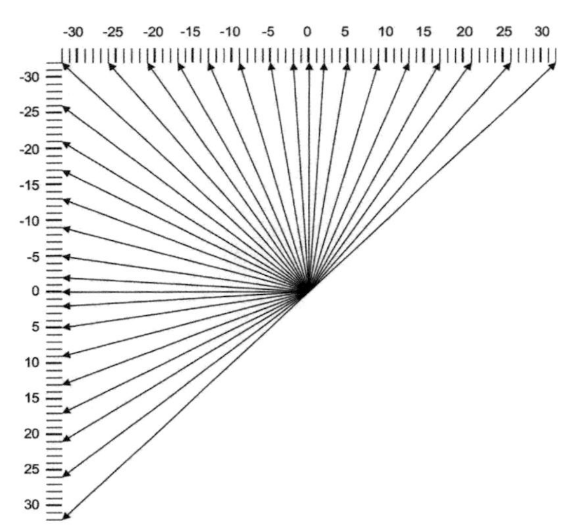

Figure 5.8 Intra prediction angle definition [E59].

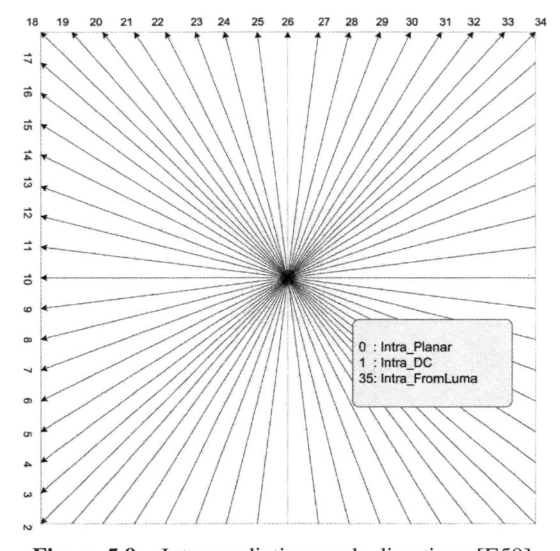

Figure 5.9 Intra prediction mode directions [E59].

Table 5.3 Mapping between intra-prediction mode direction (shown in Figure 5.9) and intra prediction angles (shown in Figure 5.8) [E125]

predModeIntra	**1**	**2**	**3**	**4**	**5**	**6**	**7**	**8**	**9**	**10**	**11**	**12**	**13**	**14**	**15**	**16**	**17**
intraPredAngle	–	32	26	21	17	13	9	5	2	0	–2	–5	–9	–13	–17	–21	–26
predModeIntra	**18**	**19**	**20**	**21**	**22**	**23**	**24**	**25**	**26**	**27**	**28**	**29**	**30**	**31**	**32**	**33**	**34**
intraPredAngle	–32	–26	–21	–17	–13	–9	–5	–2	0	2	5	9	13	17	21	26	32

Both visual and textual description of these directional modes, besides filtering process of reference samples as predictors, post processing of predicted samples and myriad other details are described in Chapter 4 Lainema and Han, "Intra-picture prediction in HEVC" [E202].

5.4.2 Transform Coefficient Scanning

The three transform coefficient scanning methods, diagonal, HOR, and vertical adopted in HEVC for a 8 × 8 transform block (TB) are shown in Figure 5.10 [E74]. The scan in a 4 × 4 transform block is diagonal. HOR and vertical scans may also be applied in the intra case for 4 × 4 and 8 × 8 transform blocks.

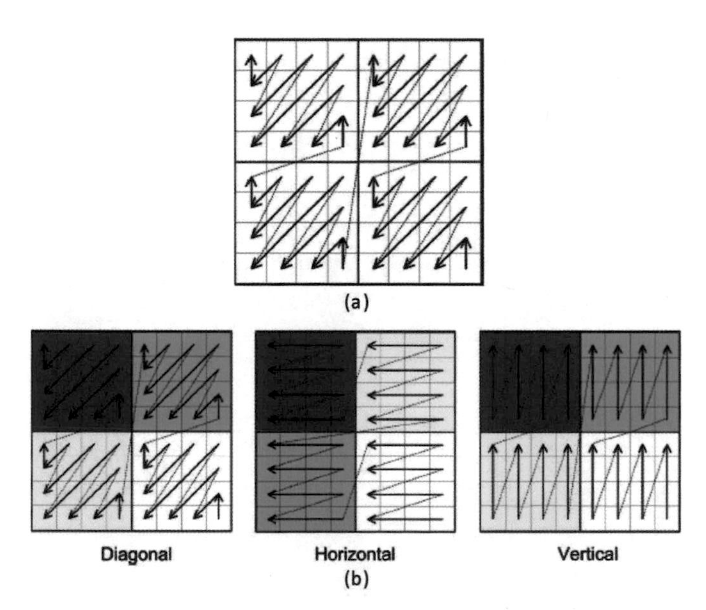

Figure 5.10 (a) Diagonal scan pattern in 8 × 8 TB. The diagonal scan of a 4 × 4 TB is used within each 4 × 4 sub-block of larger blocks and (b) coefficient groups for 8 × 8 TB [E74]. © 2012 IEEE.

Chapter 6, Budagavi, Fuldseth and Bjontegaard, "HEVC transform and quantization", in [E202] describe the 4 × 4 to 32 × 32 integer 2-D DCTs including the embedding process (small size transforms are embedded in large size transforms) default quantization matrices for transform block sizes of 4 × 4 and 8 × 8 and other details are addressed. The embedding feature allows for different transform sizes to be implemented using the same architecture thereby facilitating hardware sharing. An extensive list of references related to integer DCT architectures is provided in [E381].

An alternate 4 × 4 integer transform derived from DST for intra 4 × 4 luma blocks is also listed.

5.4.3 Luma and Chroma Fractional Pixel Interpolation

Integer (A_{i-j}) and fractional pixel positions (lower case letters) for luma interpolation are shown in Figure 5.11 [E59]. See [E69] for generalized interpolation.

$A_{-1,-1}$				$A_{0,-1}$	$a_{0,-1}$	$b_{0,-1}$	$c_{0,-1}$	$A_{1,-1}$				$A_{2,-1}$
$A_{-1,0}$				$A_{0,0}$	$a_{0,0}$	$b_{0,0}$	$c_{0,0}$	$A_{1,0}$				$A_{2,0}$
$d_{-1,0}$				$d_{0,0}$	$e_{0,0}$	$f_{0,0}$	$g_{0,0}$	$d_{1,0}$				$d_{2,0}$
$h_{-1,0}$				$h_{0,0}$	$i_{0,0}$	$j_{0,0}$	$k_{0,0}$	$h_{1,0}$				$h_{2,0}$
$n_{-1,0}$				$n_{0,0}$	$p_{0,0}$	$q_{0,0}$	$r_{0,0}$	$n_{1,0}$				$n_{2,0}$
$A_{-1,1}$				$A_{0,1}$	$a_{0,1}$	$b_{0,1}$	$c_{0,1}$	$A_{1,1}$				$A_{2,1}$
$A_{-1,2}$				$A_{0,2}$	$a_{0,2}$	$b_{0,2}$	$c_{0,2}$	$A_{1,2}$				$A_{2,2}$

Figure 5.11 Integer and fractional positions for luma interpolation [E59].

Unlike a two-stage interpolation process adopted in H.264, HEVC uses separable 8-tap filter for 1/2 pixels and 7-tap filter for 1/4 pixels (Table 5.4) [E59, E71, E109]. Similarly 4-tap filter coefficients for chroma fractional (1/8 accuracy) pixel interpolation are listed in Table 5.5. Lv et al. [E109] have conducted a detailed study of performance comparison of fractional-pel interpolation filters in HEVC and H.264/AVC and conclude that the filters in HEVC increase the BD rates [E81, E82, E198] by more than 10% compared to those in H.264/AVC at the cost of increased implementation complexity.

Table 5.4 Filter coefficients for luma fractional sample interpolation [E59]

index	–3	–2	–1	0	1	2	3	4
hfilter[i]	–1	4	–11	40	40	–11	4	1
qfilter[i]	–1	4	–10	58	17	–5	1	

Table 5.5 Filter coefficients for chroma fractional sample interpolation [E59]

index	–1	0	1	2
filterl[*i*]	–2	58	10	–2
filtei2[*i*]	–4	54	16	–2
filter3[*i*]	–6	46	28	–4
filter4[*i*]	–4	36	36	–4

M. Wien, "High Efficiency Video Coding: Coding Tools and Specification", Springer, 2014. See Section 7.3.3.

Determination of the interpolation filter (IF) coefficients pages 194–196 for the theory behind how these IF coefficients are derived using DCT.

5.4.4 Comparison of Coding Tools of HM1 and HEVC Draft 9

Coding tools of HEVC test model version 1 (HM1) and draft 9 [E59] are summarized in Table 5.6 [E59]. The overview paper on HEVC by Sullivan et al. [E61] is an excellent resource which clarifies not only all the inherent functionalities but also addresses the history and standardization process leading to this most efficient standard. In the long run, the HEVC (including the additions/extensions/profiles) has the potential/prospects/promise to overtake all the previous standards including H.264/AVC.

Table 5.6 Summary of coding tools of high-efficiency configuration in HM1 and HEVC [E59]

Functionality	HM 1 High Efficiency	(HIVC draft 9)
CIU structure	Tree structure from 8 × 8 to 64 × 64	Tree structure from 8 × 8 & to 64 × 64
PU structure	Square and symmetric	Square, symmetric and asymmetric (only square for intra)
TXJ structure	Tree structure of square TUs	Tree structure of square TUs
Core transform	Integer transforms from 4 to 32 points (full factorization)	Integer transforms from 4 to 32 points [partially factorable)
Alternative transform	n/a	Integer DST tyrje for 4 × 4
Intra prediction	33 angular modes with DC mode	33 angular modes with planar and DC modes
Luma interpolation	12-tap separable	S-tap/7-tap separable
Chroma interpolation	Bilinear	4-tap separable
MV prediction	AMVP	AMVP
MC region merging	Spatial CU merging	PU merging
Entropy coding	CABAC	CABAC
Deblocking filter	Nonparallel	Parallel
Sample adaptive offset	n/a	Enabled
Adaptive loop filter	Multiple shapes	n/a
Dedicated tools for parallel processing	Slices	Slices,, tiles, wavefronts and dependent slice segments

5.5 Extensions to HEVC

As with H.264/AVC, additions/extensions to HEVC include 4:2:2 and 4:4:4 formats, higher bit depths (10 and 12), scalable video coding (SVC), (References are listed in Section 5.10 Summary) 3D/stereo/multiview coding and screen content coding (See [E23, E25, E34, E39, E260, E326, SCC48]). Several proposals related to SVC have been presented at the Shanghai meeting of HEVC (ITU-T/ISO, IEC) held in Oct. 2012. The reader is referred to the poster session [E27] MA.P2 "High efficiency video coding", IEEE ICIP 2012, Orlando FL, Sept.–Oct. 2012 and the special issue on emerging research and standards in next generation video coding" *IEEE Trans. CSVT*, vol. 22, Dec. 2012 [E45]. Several papers from this special issue are cited as references here [E61–E63, E67, E73, E74, E76-E79, E83–E85, E87–E93, E120], see also [E321, E325, E326].

These papers cover not only the various aspects of HEVC, but also ways to reduce the implementation complexity and the reasons behind the various tools and techniques adopted in this standard. The justification for initial consideration of some tools such as adaptive loop filter, MDDT, ROT, etc. and subsequent elimination of these tools in the standard is also provided in some of these papers.

Some of these tools are reconsidered in the NGVC, beyond HEVC. Also IEEE Journal of selected topics of signal processing has published a special issue on video coding: HEVC and beyond, Vol. 7, December 2013 [E80].

These additions/extensions have been standardized in 2015.

5.6 Profiles and Levels

Three profiles (Main, Main10 and Main Still Picture – intra-frame only) are listed in Annex A in the final draft international standard (FDIS) (January 2013) [E125]. ITU-T study group 16 has agreed to this first stage approval formally known as Recommendation H.265 or ISO/IEC 23008-2. Main profile is limited to YC_bC_r 4:2:0 format, 8-bit depth, progressive scanning (non-interlaced) with spatial resolutions ranging from QCIF (176 $\times$ 144) to 7640 $\times$ 4320 (called 8K $\times$ 4K). Figure 5.12 [E23] lists the spatial resolutions ranging from SD (NTSC) to super Hi-Vision/ultra HD video. In the main profile 13 levels are included in the first version (Table 5.7) [E59].

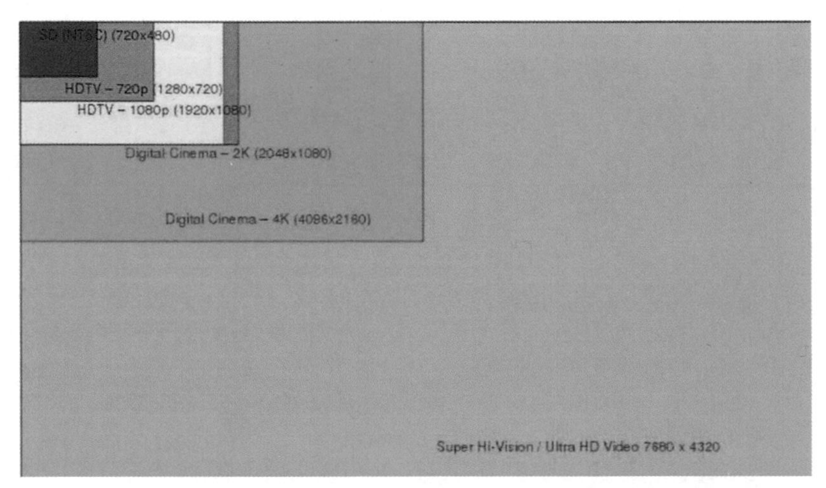

Figure 5.12 Future visual applications and demands (NHK and Mitsubishi have developed a SHV-HEVC hardware encoder).

Table 5.7 Level limits for the main profile in HEVC [E59]

Level	Max Luma Picture Size (Samples)	Max Luma Sample Rate (Samples/Sec)	Main Tier Max Bit Rate (1000 bits/s)	High Tier Max Bit Rate (1000 bits/s)	Min Comp. Ratio
1	36,864	552,960	128	–	2
2	122,830	3,686,400	1,500	–	2
2.1	245,760	7,372,800	3,000	–	2
3	552,960	16,588,800	6,000	–	2
3.1	983,040	33,177,600	10,000	–	2
4	2,228,224	66,846,720	12,000	30,000	4
4.1	2,228,224	133,693,440	20,000	50,000	4
5	8,912,896	267,386,830	25,000	100,000	6
5.1	8,912,396	534,773,760	40,000	160,000	8
5.2	8,912,396	1,069,547,520	60,000	240,000	8
6	33,423,360	1,069,547,520	60,000	240,000	8
6.1	33,423,360	2,005,401,600	120,000	480,000	8
6.2	33,423,360	4,010,803,200	240,000	800,000	6

5.7 Performance and Computational Complexity of HEVC Encoders

Correa et al. [E84] have carried out a thorough and detailed investigation of the performance of coding efficiency versus computational complexity of HEVC encoders. This investigation focuses on identifying the tools that most affect these two vital parameters (efficiency and complexity). An invaluable outcome is the tradeoff between the efficiency and complexity useful for implementing complexity-constrained encoders.

Additionally, the development of low-complexity encoders that achieve coding efficiency comparable to high-complexity configurations is a valuable resource for industry. This combination of low-complexity and high efficiency can be achieved by enabling Hadamard ME, asymmetric motion partitions and the loop filters instead of computationally demanding tools such as ME. Their analysis included the three coding tools (non-square transform, adaptive loop filter and LM chroma) which have been subsequently dropped from the draft standard [E60].

5.8 System Layer Integration of HEVC

Schierl et al. describe the system layer integration of HEVC [E83]. The integration of HEVC into end-to-end multimedia systems, formats, and protocols (RTP, MPEG-2 TS, ISO File Format, and dynamic adaptive streaming over HTTP - DASH) is discussed. Error resiliency tools in HEVC are also addressed. Many error resilience tools of H.264/AVC such as FMO, ASO, redundant slices, data partitioning and SP/SI pictures (Chapter 4) have been removed due to their usage. They suggest that the use of error concealment in HEVC should be carefully considered in implementation and is a topic for further research. They conclude with a description of video transport and delivery such as broadcast, IPTV, internet streaming, video conversation and storage as provided by the different system layers. See reference 24 listed in this paper, i.e., T. Schierl, K. Gruenberg and S. Narasimhan, "Working draft 1.0 for transport of HEVC over MPEG-2 Systems", ISO/IEC SC29/WG11, MPEG99/N12466, February 2012. See other references related to various transport protocols.

5.9 HEVC Lossless Coding and Improvements [E85]

Zhou et al. [E85] have implemented lossless coding mode of HEVC main profile (bypassing transform, quantization, and in-loop filters – Figure 5.13 [E85]) and have shown significant improvements (bit rate reduction) over current lossless techniques such as JPEG2000, JPEG-LS, 7-Zip and WinRAR. They improve the coding efficiency further by introducing the sample based intra angular prediction (SAP).

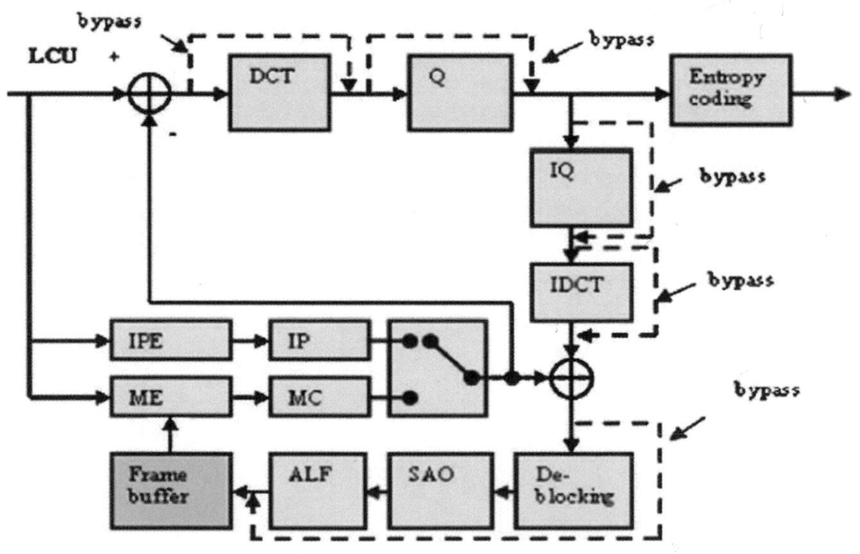

Figure 5.13 HEVC encoder block diagram with lossless coding mode [E85] (See Figure 5.5) © 2012 IEEE.

Cai et al. [E114] also conducted the lossless coding of HEVC intra mode, H.264 High 4:4:4 Profile intra mode, MJPEG 2000 and JPEG-LS using a set of recommended video sequences during the development of HEVC standard. Their conclusion is that the performance of HEVC has matched that of H.264/AVC and is comparable to JPEG-LS and MJPEG 2000. Similar tests on lossy intra-coding show that HEVC high10, H.264/AVC HP 4:4:4 and HEVC MP have similar performance. However, MJPEG 2000 outperforms the former three in low bit scenario, while this advantage is gradually compensated and finally surpassed by the former three as the bit rate increases. Several other interesting papers on performance comparison of these and other standards are listed in [E114]. Wige et al. [E147] have implemented HEVC lossless coding using pixel-based averaging predictor. Heindel, Wige and Kaup [E194] have developed lossless compression of enhancement layer (SELC) with base layer as lossy. Wang et al. [E196] have implemented lossy to lossless image compression based on reversible integer DCT.

Horowitz et al. [E66] have conducted the informal subjective quality (double blind fashion) comparison of HEVC MP (reference software HM 7.1) and H.264/AVC HP reference encoder (JM 18.3) for low delay applications. Table 5.8 describes the comparison results for test sequences encoded using the HM and JM encoders. They conclude that HEVC generally produced better subjective quality compared with H.264/AVC at roughly half the bit rate with viewers favoring HEVC in 73.6% trials.

Table 5.8 Subjective viewing comparison results for sequences encoded using the HM and JM encoders [E66] © SPIE 2012

Subjects Viewing for HM vs. JM									
Sequence	HM Bit Rate (kbps)	HM QP	JM Bit Rate (kbps)	JM QP	HM: JM Bit Rate	Votes Favoring HM	Vote; Favoring JM	% Favoring HEVC	Total Number of Votes
KristenAndSara	149	38	302	37	49%	10	15	40%	25
Vidyol	190	37	367	*36*	52%	14	11	56%	25
OldTownCross	408	37	879	37	46%	22	3	88%	25
Kimono 1	682	36	1404	35	49%	21	4	84%	25
toys_and_calendar	347	37	734	38	47%	25	0	100%	25
				Average	**–49%**	92	33	**73.6%**	125

To reinforce these results, x264 the production quality H.264/AVC encoder is compared with eBrisk Video (production quality HEVC implementation), see Table 5.9 for the comparison results. Viewers favored eBrisk-encoded video at roughly half the bit rate compared to x264 in 62.4% trials. These tests confirm that HEVC yields similar subjective quality at half the bit rate of H.264/AVC. Chapter 9—Compression performance analysis in HEVC—by Tabatabai et al. in [E202] describes in detail the subjective and objective quality comparison of HEVC with H.264/AVC. Tan et al. [4] have presented the subjective and objective comparison results of HEVC with H.264/AVC and conclude that the key objective of the HEVC has been achieved i.e., substantial improvement in compression efficiency compared to H.264/AVC.

Table 5.9 Subjective viewing comparison results for sequences encoded using the eBrisk and x264 encoders [E66] © 2012 SPIE

Subjective Viewing for eBrisk vs. x264									
Sequence	eBiisk Bit Rate (kbps)	eBrisk QP	x564 Bit Rate (kbps)	x264 QP	eBrisk: x264 Bit Rate	Votes Favoring eBrisk	Votes Favoring x264	% Favoring HEVC	Total Number of Votes
KristenAndSara	332	36	657	33	51%	12	13	48%	25
Vidvol	363	36	773	32	47%	10	15	40%	25
OldTownCross	904	35	1716	34	53%	22	3	88%	25
Kimono 1	1384	35	2670	32	52%	17	8	68%	25
toys_and_calendar	729	36	1553	33	47%	17	8	68%	25
				Average	**50%**	78	47	**62.4%**	125

5.10 Summary

In the family of video-coding standards (Figure 5.1), HEVC has the promise and potential to replace/supplement all the existing standards (MPEG and H.26x series including H.264/AVC). While the complexity of the HEVC encoder is several times that of the H.264/AVC, the decoder complexity [E163] is within the range of the latter. Researchers are exploring about reducing the HEVC encoder complexity [E63, E79, E84, E88, E89, E106, E108, E166 - E17, E187, E195R, E199, E204, E205, E208, E221, E230, E232, E240, E242, E247, E248, E252, E257, E258, E259, E269, E270, 274, E276, E278, E290, E308, E312, E332, E348, E368, E369, E374, E383, E388, E390].

Bin and Cheung [E247] have developed a fast decision algorithm for the HEVC Intra coder and report encoding time is reduced by 52% on average compared with HM 10.0 reference software with negligible losses. Similarly, Kim and Park [E248] developed a fast CU partitioning algorithm resulting in reducing the computational complexity of the HEVC encoder by 53.6% on the average with negligible BD bit rate loss. Chi et al. [E246] have developed a SIMD optimization for the entire HEVC decoder resulting in up to 5x speed up. Min, Xu and Cheung [E382] have developed a fully pipelined architecture for intra prediction of HEVC that achieves a throughput of 4 pixels in one clock cycle and is capable to support 3840 × 2160 video at 30 fps. Correa et al. [E383] have developed techniques that result in significant decrease in encoding time with negligible loss in compression efficiency. (part of the abstract is repeated here).

> "**A** set of Pareto-efficient encoding configurations, identified through rate-distortion-complexity analysis. This method combines a medium-granularity encoding time control with a fine-granularity encoding time control to accurately limit the HEVC encoding time below a predefined target for each GOP. The encoding time can be kept below a desired target for a wide range of encoding time reductions, e.g., up to 90% in comparison with the original encoder. The results also show that compression efficiency loss (Bjøntegaard delta-rate) varies from negligible (0.16%) to moderate (9.83%) in the extreme case of 90% computational complexity reduction."

In particular, refer to [E242], wherein significant encoder complexity reduction is achieved with negligible increase in BD-rate by combining early termination processes in CTU, PU and RQT structures. See also P.5.139 related to this. Chapter 4—Intra-picture prediction in HEVC—and Chapter 5—inter-picture prediction in HEVC—and the references at the end of these chapters [E202] provide insights related to encoder complexity reduction. Another approach in reducing the encoder complexity is based on zero block detection [E166, E171, E186]. Kim et al. [E168] have shown that motion estimation (ME) occupies 77–81% of HEVC encoder implementation. Hence, the focus has been in reducing the ME complexity [E137, E152, E168, E170, E179, E205, E221, E239, E275, E351, E359, E370, E371]. PPT slides related to fast ME algorithm and architecture design are presented in [Tut10]. Machine learning techniques [E259, E263, E317, E359] have been applied effectively resulting in significant reduction of HEVC encoder complexity.

Several researchers have implemented performance comparison of HEVC with other standards such as H.264/AVC (Chapter 4 in [Book4]), MPEG-4 Part 2 visual, H.262/MPEG-2 Video [E183], H.263, and VP9 (Chapter 6 in [Book4]), and also with image coding standards such as JPEG2000, JPEG-LS, and JPEG-XR [E22, E47, E62, E63, E65, E66, E114, E177, E191, E245]. Similar comparisons with MPEG-4 Part 10 (AVC) and VP9 are reported in [E191]. See also papers listed under VP8, VP, and VP10. Most significant of all these is the detailed performance comparison (both objective and subjective) described in [E62]. Special attention may be drawn to references 39, 40, 42–47 cited in [E62]. Chapter 9—Comparison performance analysis in HEVC—in [E202] provides both objective/subjective comparison between HM 12.1 (HEVC) and JM 18.5 (AVC – H.264). See Table 9.4 in [E202] for description of the test conditions.

However, the payoff is several tests have shown that HEVC provides improved compression efficiency up to 50% bit rate reduction [E61, E62, E99] for the same subjective video quality compared to H.264/AVC (Chapter 4 in [Book4]). See Tables 5.8 and 5.9. Besides addressing all current applications, HEVC is designed and developed to focus on two key issues: increased video resolution—up to 8k × 4k—and increased use of parallel processing architecture. A detailed description of the parallelism in HEVC is outlined in Chapter 3—H. Schwarz, T. Schierl and D. Marpe, "Block structures and parallelism features in HEVC", in [E202]. This chapter also demonstrates that more than half of the average bit-rate savings compared to H.264 are due to increased flexibility of block partitioning for prediction and transform coding. Brief description of the HEVC is provided. However, for details and implementation, the reader is referred to the JCT-VC documents [E55, E188], overview papers (See section on overview papers).

[E321] is an overview paper describing the range extensions (Rext) of the HEVC. Beyond the Main, Main 10 and Main Intra profiles finalized in version1, Rext (version 2) focuses on monochrome, 4:2:4 and 4:4:4 sampling formats and higher bit depths. New coding tools integrated into version 2 are: Cross-component prediction (CCP) [E355], ACQP and RD-PCM. Modification to HEVC Version 1 tools are: Filtering for smoothing of samples in intra-picture prediction, transform skip and transform quantizer bypass modes, truncated Rice binarization, internal accuracy and *k*-th order Exp-Golomb (EGk) binarization, and CABAC bypass decoding. These tools and modifications are meant to address the application scenarios such as: content production in a digital video broadcasting delivery chain, storage, and transmission of video captured by: a professional camera, compression of high dynamic range content, improved lossless compression, and coding of screen content. This overview paper describes in detail not only the justification for introducing these additional tools, modifications and profiles but also the improved performance. Using a spectrum of standard video test sequences the authors have compared the performance of HEVC version II (HM 16.2 software) with H.264/AVC (JM 18.6) using BD bit rate as a metric. This comparison extends to lossless coding for different temporal structures called as AI, RA, and LD. In the overview paper on SCC [E322], implementation complexity as a performance measure is conspicuous by its absence. Extensive list of references related to SCC are cited at the end. See also the overview paper on SCC [SCC48].

Keynote speeches [E23, E99], tutorials (See section on tutorials), panel discussions [E98, E366, E380], poster sessions [E27, E139], workshop [E257] special issues (See special issues on HEVC) test models (TM/HM) [E54], web/ftp sites [E48, E53–E55], open source software (See section on open source software), test sequences, file formats [E327] and anchor bit streams [E49]. Also researchers are exploring transcoding between HEVC and other standards such as MPEG-2 (See section on transcoders). Further extensions to HEVC [E158] are scalable video coding (SHVC) [E76, E119, E126–E129, E141, E158, E159, E208, E213, E214, E234, E238, E264, E284, E325] (note that E325 is an overview paper on SHVC), 3D video/multiview video coding [E34, E39, E201, E227, E271, E326] (note that E326 is an overview paper on the 3D video/multiview coding) and range extensions [E155, E321] which include screen content coding, See the references on screen content coding besides the overview paper [E322] bit depths larger than 10 bits and color sampling of 4:2:2 and 4:4:4 formats.

In an overview paper, [E325] Boyce et al. describe the scalable extensions of the HEVC (SHVC), which include temporal, spatial, quality (SNR), bit depth and color gamut functionalities and as well their combination (hybrid). They compare the performance of SHVC with the scalable extensions of H.264/AVC called SVC and conclude using various test sequences (Table 5 in [E325]), that SHVC is 50–63% more efficient than SVC depending on the scalability type and the use scenario. In view of the superior performance, various standardization bodies such as 3GPP and ATSC for broadcasting and video streaming and IMTC for videoconferencing are evaluating SHVC as potential codec choice. Investigation of bit rate allocation in SHVC is addressed in [E350].

[E322, SCC48] are overview papers on the HEVC SCC extension finalized in 2016. They describe several new modules/tools added to the original

HEVC framework and also the new coding tools introduced during the development of HEVC–SCC. [E322] concludes with performance analysis of the new coding tools relative to HEVC—Rext and H.264/AVC.

Screen content coding (SCC) [E322, SCC48] in general refers to computer generated objects and screen shots from computer applications (both images and videos) and may require lossless coding. HEVC-SCC extension reference software is listed in [E256]. Some aspects of SCC are addressed in [E352, E353]. They also provide fertile ground for R & D. Screen content coding has been finalized in 2016. [SCC15] is the special issue on screen content video coding and applications published in Dec. 2016 and reports briefly on ongoing standardization activities. This special issue covers on all aspects of SCC besides opening the avenue for a number of projects/problems. The overview paper on screen content video coding (SCVC) by Peng et al. [OP11] addresses recent advances in SCC with an emphasis on two state-of-the-art standards: HEVC/H.265 Screen Content Coding Extensions (HEVC-SCC) by ISO/IEC MPEG and Display Stream Compression (DSC) by VESA. They also explore future research opportunities and standards developments in order to give a comprehensive overview of SCVC. Iguchi et al. [E164] have already developed a hardware encoder for super hi-vision (SHV) i.e., ultra HDTV at 7680 × 4320 pixel resolution. Shishikui [E320] has described the development of SHV broadcasting in Japan and states that 8K SHV broadcasting to begin in 2016 during the Rio de Janeiro Summer Olympics and for full pledged broadcasting to begin in 2020 with the Tokyo Summer Olympics. He has also presented a keynote speech “8K super hi-vision: science, engineering, deployment and expectations”, in IEEE PCS 2015 in Cairns, Australia.

Tsai et al. [E222] have designed a 1062 Mpixels/s 8192 × 4320 HEVC encoder chip. See also [E165] which describes real-time hardware implementation of HEVC encoder for 1080p HD video. NHK is planning SHV experimental broadcasting in 2016. See also [E226]. A 249-Mpixel/s HEVC video decoder chip for 4k Ultra-HD applications has already been developed [E182]. Bross et al. [E184] have shown that real time software decoding of 4K (3840 × 2160) video with HEVC is feasible on current desktop CPUs using four CPU cores. They also state that encoding 4K video in real time on the other hand is a challenge. AceThought among many other companies provides high-quality and efficient software implementation of HEVC Main Profile, Main 10 Profile and Main Still Picture Profile decoder algorithms on various handheld and desktop platforms [E189]. Cho et al. [E219] have developed CTU level parallel processing for single core and picture partition level multi core parallel processing for 4K/8K UHD high performance HEVC hardware decoder implementation. They conclude that their multi core decoder design is suitable for 4K/8K UHD real time decoding of HEVC bit streams when implemented on SoC. Kim et al. [E220] have extended this to scalable HEVC hardware decoder and the decoder design is estimated to be able to decode 4K@60 fps HEVC video. Decoder hardware architecture and encoder hardware architecture for HEVC are addressed in Chapters 10 and 11, respectively, in the book Sze, Budagavi and Sullivan (editors) "High efficiency video coding (HEVC): algorithms and architectures", Springer, 2014 [E202]. For 4K video, software decoder has been developed [E329]. A hardware architecture of a CABAC decoder for HEVC that can lead to real time decoding of video bit streams at levels 6.2 (8K UHD at 10 fps) and 5.1 (4K UHD at 60 fps) has been developed by Chen and Sze [E249]. See also [E390]. Zhou et al. also have developed similar VLSI architecture for H.265/HEVC CABAC encoder [E250]. Wang et al. [E359] have presented an improved and highly efficient VLSI implementation of HEVC MC that supports 7680 $\times$ 4320@60fps bidirectional prediction. The authors [E359] claim that the proposed MC design not only achieves 8x throughput enhancement but also improves hardware efficiency by at least 2.01 times in comparison with prior arts. Zooming and displaying region of interest (ROI) from a high resolution video i.e., ROI based streaming in HEVC has also been developed by Umezaki and Goto [E205]. The authors suggest future work to reduce the decoding cost. A novel Wavefront-Based High parallel Solution for HEVC Encoding integrating data level, optimal single Instruction – Multiple – Data (SIMD) algorithms are designed by Chen et al. [E272], in which the overall WHP solution can bring up to 57.65x, 65.55x and 88.17x speedup for HEVC encoding of WVGA, 720p and 1080p standard test sequences while maintaining the same coding performance as with Wavefront Parallel Processing (WPP). Abeydeera et al. [E338] have developed a 4K real time decoder on FPGA. Chiang et al. [E345] have developed a real-time HEVC decoder capable of decoding 4096 $\times$ 2160@30fps at 270 MHz. Open source code for HEVC encoder is developed by kvazaar [0SS22]. Transport of HEVC video over MPEG-2 systems is described in [E197]. Encode/decode and multiplex/demultiplex HEVC video/AAC audio while maintaining lip synch has been implemented in software [E203]. This, however, is limited to HEVC main profile intra coding only. Extension to inter prediction and also to other profiles are valid research areas. Kim, Ho and Kim [E254] have developed a HEVC-complaint perceptual video coding scheme based on

JND models that has significant bit-rate reductions with negligible subjective quality loss at a slight increase in encoder complexity. Several companies and research institutes have developed various products related to HEVC (see the booths and exhibits at NAB 2016, held in Las Vegas 16–21 April 2016 www.nabshow.com). One highlight is the single chip #8K #video decoder developed by Socionext US. Several companies have developed or developing hardware encoders/decoders (chips or SOC) that can process up to 4 and 8K video [TUT10]. Details on software codecs are also provided. This tutorial concludes with future challenges for next generation video encoders that relate to extensions such as HSVC, SCC, Multiview, 3D etc. A new standard called MPEG-4 internet video coding (IVC) has been approved as IS0/IEC 14496-33 in June 2015 [E399]. This standard is highly beneficial for video services in the internet domain. The HEVC standard and all its extensions are expected to go through the various revisions and refinements meeting the growth in consumer electronics, internet, VLSI, security, defense, medical, broadcasting, and related areas over the next several years. The next major standard is projected to be around 2020 designed to meet 5G specification. Chen et al. [BH1] and Alshin et al. [BH2] address video coding techniques beyond HEVC. References [1] through [8] in Alshin et al. [BH2] describe the algorithms and techniques proposed for NGVC beyond HEVC. Sze and Budagavi and also Grois et al. (see tutorials) have suggested directions for post HEVC activities. While the main emphasis in developing HEVC has been on efficient video compression and coding, it is interesting to note that the features in its compressed domain have been adopted in video saliency detection [E400].

The projects listed below range from graduate 3 credit hours to research at the M.S. and doctoral levels. These projects are designed to provide additional insights into the tools and techniques and provide a forum for their modifications leading to further improvements in the HEVC encoder/decoder.). Besides the references related to HEVC, references are listed under various categories such as, books, tutorials, transcoders, on line courses (OLC), overview papers, open source software, video test sequences, SSIM, screen content coding, file format, JVT reflector, MSU codec comparison, general, beyond HEVC etc. for easy access. References and projects related to JPEG, JPEG-LS, JPEG-2000, JPEG-XR, JPEG XS, JPEG-XT, PNG, WebP, WebM, DIRAC and DAALA are added at the end. Keynote address by T. Ebrahimi "JPEG PLENO: Towards a new standard for plenoptic image compression", IEEE DCC, Snow Bird, Utah,

2016 along with the abstract is included. This developing new standard while maintaining backward compatibility with legacy JPEG opens up thought provoking and challenging research [JP-P1]. Panel presentations by Sullivan and others [E366] "Video coding: Recent developments for HEVC and future trends' during IEEE DCC 2016 opens up research areas for next generation video coding (NGVC). Slides related to high dynamic range, wide color gamut video, 360 video, royalty free video standards are of special interest providing fertile ground for research. Access to HEVC SCC extensions [OP11], new profiles, further compression exploration beyond HEVC is also provided. Several tests [E372] have shown superior performance of HEVC intra compared to JPEG, JPEG LS, JPEG 2000, JPEG XT, JPEG XR and VP9 intra in terms of PSNR and SSIM. Related problems are described in P.5.237 and P.5.238. However, implementation complexity of HEVC intra is higher compared to all the others. Brief introduction to AV1 codec being developed by Alliance for Open Media (AOM) is included. This is followed by brief description of Real Media HD (RMHD) codec developed by Real Networks. Special sessions on recent developments in video coding, genomic compression and quality metrics/perceptual compression to be held in IEEE DCC, April 2017 are of interests to researchers, academia and industry [E386]. Projects related to demosaicking of mosaicked images are listed in P.5.254 thru P.5.257 (See [E384] and [E387]). In IEEE DCC 2017 [E386] special sessions on "Video coding", "Genome Compression", and "Quality Metrics and Perceptual Compression" can lead to number of R&D projects. Of particular interest is perceptual compression which is based on psycho visual perception (vision science). Another interesting topic is keynote address "Video Quality Metrics" by Scott Daly, Senior Member Technical Staff, Dolby Laboratories. Besides compression and coding, some features of HEVC in compressed domain are used in developing video saliency detection [E400]. The authors in [E400] claim that their method is superior to other state-of-the-art saliency detection techniques. Recently, Kim et al. [E355] have demonstrated significant coding gains for both natural and screen content video in HEVC based on cross component prediction (CCP). IEEE ICCE, Las Vegas, Jan. 2017 has also many papers related to HEVC. The domain of the video coding standards, in general, is limited to encoder/decoder. Both preprocessing (RGB to color space conversion, quantization and chroma subsampling) and post processing (corresponding inverse processes) are trends for future video codecs. For example HDR and WCG include both pre and post processing. This implies end-to-end pipeline video capture and display in RGB format.

5.11 Projects

There are some projects which refer to EE Dept., University of Texas at Arlington (UTA), Arlington, Texas, 76019, USA. For details please go to **www.uta.edu/faculty/krrao/dip**. Click on courses and then click on Graduate courses followed by EE5359 Multimedia Processing. Scroll down to theses and also to projects.

P.5.1 Deng et al. [E17] have added further extensions to H.264/AVC FRExt such as larger MV search range, larger macroblock, skipped block sizes and 1-D DDCT. They compared its performance with motion JPEG 2000 using high resolution (HR) (4096 $\times$ 2160) video sequences and showed significant improvement of the former in terms of PSNR at various bit rates. Implement the extended H.264/AVC and Motion JPEG 2000 and confirm that the former has a superior performance using HR test sequences. C. Deng et al., "Performance analysis, parameter selection and extension to H.264/AVC FRExt for high resolution video coding", J. VCIR, vol. 22, pp. 687–760, Feb. 2011.

P.5.2 Karczewicz et al. [E10] have proposed a hybrid video codec superior to H.264/AVC codec by adding additional features such as extended block sizes (up to 64 $\times$ 64), mode-dependent directional transforms (MDDT) for intra-coding, luma and chroma high-precision filtering, adaptive coefficient scanning, extended block size partition, adaptive loop filtering, large size integer transform, etc. By using several test sequences at different spatial resolutions, they have shown that the new codec out performs the traditional H.264/AVC codec in terms of both subjective quality and objective metrics. Also this requires only moderate increase in complexity of both the encoder and decoder. Implement this new codec and obtain results similar to those described in this paper, consider SSIM [SSIM1] also as another metric in all the simulations. Use the latest JM software for H.264/AVC. M. Karczewicz et al., "A hybrid video coder based on extended macroblock sizes, improved interpolation, and flexible motion representation", IEEE Trans. CSVT, vol. 20, pp. 1698–1708, Dec. 2010.

P.5.3 Ma and Segall [E20] have developed a low resolution (LR) decoder for HEVC. The objective here is to provide a low power decoder within a high-resolution bit stream for handheld and mobile devices. This is facilitated by adopting hybrid frame buffer compression,

LR intra prediction, cascaded motion compensation, and in loop deblocking [E73], within the HEVC framework. Implement this low power HEVC decoder. Also port these tools in the HEVC reference model (HM9.0) [E56] and evaluate the performance. Z. Ma and A. Segall, "Low resolution decoding for high-efficiency video coding", IASTED SIP 2011, pp., Dallas, TX, Dec. 2011.

P.5.4 Joshi, Reznik and Karczewicz [E8] have developed scaled integer transforms which are numerically stable, recursive in structure and are orthogonal. They have also embedded these transforms in H.265/JMKTA framework. Specifically develop the 16-point scaled transforms and implement in H.265 using JMKTA software. Develop 32 and 64 point scaled transforms. R. Joshi, Y.A. Reznik and M. Karczewicz, "Efficient large size transforms for high-performance video coding", Applications of Digital Image Processing XXXIII, Proc. of SPIE, vol. 7798, 77980W-1 through 77980W-7, 2010.

P.5.5 Please access S. Subbarayappa's thesis (2012) from EE 5359, MPL web site, "Implementation and Analysis of Directional Discrete Cosine Transform in Baseline Profile in H.264". Obtain the basis images for all the directional modes related to (4×4) and (8×8) DDCT^{+}. Modes 4, 6, 7, and 8 can be obtained from modes 3 and 5 as shown in Figures 13–16 (project). See also [E112]. Use this approach for obtaining the basis images.

$^{+}$Please access: http://www.h265.net/2009/9/mode-dependent-directional-transform-mddt-in-jmkta.html

P.5.6 Please access the web site http://www.h265.net/and go to analysis of coding tools in HEVC test model (HM 1.0) – intra-prediction. It describes that up to 34 directional prediction modes for different PUs can be used in intra prediction of H.265. Implement these modes in HM 1.0 and evaluate the H.265 performance using TMuC HEVC software [E97]. (HM: HEVC test model, TMuC – Test Model under Consideration).

P.5.7 Using TMuC HEVC software [E97], implement HM1.0 considering various test sequences at different bit rates. Compare the performance of HEVC (h265.net) with H.264/AVC (use JM software) using SSIM [SSIM1], bit rates, PSNR, BD measures [E81, E82, E96, E198] and computational time as the metrics. Use WD 8.0 [E60].

P.5.8 In the document JCTVC-G399 r2, Li has compared the compression performance of HEVC WD4 with H.264/AVC high profile. Implement this comparison using HEVC WD8 and the latest JM software for H.264/AVC based on several test sequences at different bit rates. As before, SSIM [SSIM1], PSNR, bit rates, BD measures [E81, E82, E96, E198] and implementation complexity are the metrics. JCT-VC, 7th meeting, Geneva, CH, 21–30, Nov. 2011. (comparison of compression performance of HEVC working draft 4 with H.264/AVC High profile).

P.5.9 Please access J.S. Park and T. Ogunfunmi, "A new approach for image quality assessment", ICIEA 2012, Singapore, 18–20 July 2012. They have developed a subjective measure (similar to SSIM) for evaluating video quality based on (8 × 8) 2D-DCT. They suggest that it is much simpler to implement compared to SSIM [SSIM1] while in performance it is close to SSIM. Evaluate this based on various artifacts. Also consider (4 × 4) and (16 × 16) 2D-DCTs besides the (8 × 8). Can this concept be extended to integer DCTs. Can DCT be replaced by DST (discrete sine transform).

P.5.10 Please access J. Dong and K.N. Ngan, "Adaptive pre-interpolation filter for high efficiency video coding", J. VCIR, vol. 22, pp. 697–703, Nov. 2011. Dong and Ngan [E16] have designed an adaptive pre-interpolation filter (APIF) followed by the normative interpolation filter [E69]. They have integrated the APIF into VCEG's reference software KTA 2.6 and have compared with the non-separable adaptive interpolation filter (AIF) and adaptive loop filter (ALF). Using various HD sequences, they have shown that APIF outperforms either AIF or ALF and is comparable to AIF+ALF and at much less complexity. Implement the APIF and confirm their conclusions.

P.5.11 Please access W. Ding et al., "Fast mode dependent directional transform via butterfly-style transform and integer lifting steps", J. VCIR, vol. 22, pp. 721–726, Nov. 2011 [E17]. They have developed a new design for fast MDDT through integer lifting steps. This scheme can significantly reduce the MDDCT complexity with negligible loss in coding performance. Develop the fast MDDT with integer lifting steps for (4 × 4) and (8 × 8) and compare its performance (see Figures 6–10) with the DCT and BSTM (butterfly style transform matrices) using video test sequences.

P.5.12 Please access B. Li, G.J. Sullivan and J. Xu, "Compression performance of high efficiency video coding (HEVC) working draft 4", IEEE ISCAS, pp. 886–889, Seoul, Korea, May 2012 [E22]. They have compared the performance of HEVC (WD4) with H.264/AVC (JM 18.0) using various test sequences. They have shown that WD4 provides a bit rate savings (for equal PSNR) of about 39% for random access applications, 44% for low-delay use and 25% for all intra use. Verify these tests.

P.5.13 Please access the paper E. Alshina, A. Alshin and F.C. Fernandez, "Rotational transform for image and video compression", IEEE ICIP, pp. 3689–3692, 2011. See also [BH2].

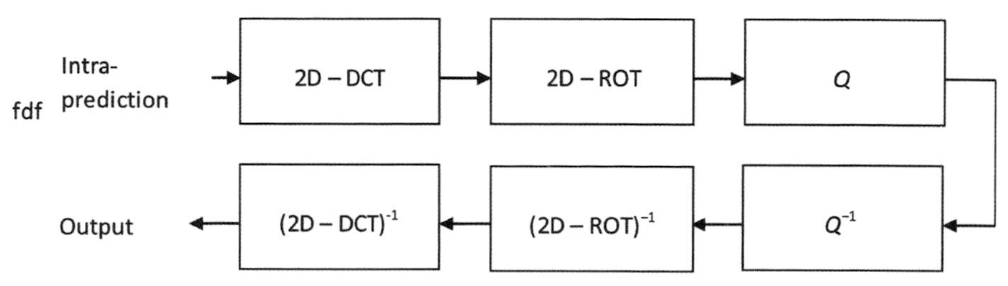

Figure 5.14 Block diagram for DCT/ROT applied to intra prediction residuals only.

Alshina, Alshin and Fernandez have applied ROT 4 to 4 × 4 blocks and ROT8 to upper left sub matrix in all other cases (see Figures 2 and 3 in the paper), and have shown a BD-rate gain of 2.5% on average for [E81, E82, E96, E198] all test sequences (see Table 4 in the paper). Implement this technique using the test sequences and confirm the results (ROT – rotational transform).

P.5.14 Please access the document JCTVC-C108, Oct. 2010 submitted by Saxena and Fernandez, (Title: Jointly optimal prediction and adaptive primary transform). They have compared TMuC 0.7 between the proposed adaptive DCT/DST as the primary transform and the DCT in intra prediction for 16 × 16, 32 × 32, and 64 × 64 block sizes for two cases i.e., secondary transform (ROT) is off or on. Implement this scheme and verify the results shown in Tables 2 and 3 of this document. Use TMuC 0.7.

P.5.15 In the Stockholm, Sweden JCT-VC meeting, adaptive DCT/DST has been dropped. Also directional DCT [E112] (to the residuals of adaptive intra directional prediction) is not considered. So also the rotational secondary transform (See P.5.13). Only a transform derived from DST for 4 × 4 size luma intra prediction residuals and integer DCT for all other cases (both intra and inter) have been adopted. The DDCT and ROT (rotational transform) contribute very little to image quality but at the cost of significant increase in implementation complexity.

See the paper by A. Saxena and F.C. Fernandez, "On secondary transforms for prediction residuals", IEEE ICIP 2012, Orlando, FL, 2012 [E26]. They have implemented the HEVC using mode dependent DCT/DST to (4 × 4) sizes for both intra and inter prediction residuals. For all other cases, (i.e., both intra and inter block sizes other than 4 × 4), they have applied a secondary transform to the top left (low frequency) coefficients after the primary 2D-DCT. This has resulted in BD rate gains (Tables 1–3) [E81, E82, E96, E198] for various test sequences compared to the case where no secondary transform is implemented. Implement this scheme and show results similar to Tables 1–3.

P.5.16 Please access H. Zhang and Z. Ma, "Fast intra prediction for high efficiency video coding", Pacific Rim Conf. on Multimedia, PCM 2012, Singapore, Dec. 2012 [E44], (http://cement.ntu.edu.sg/pcm2012/index.html)

Zhang and Ma [E44] (see also [E149] have proposed a novel intra prediction approach at the PU level and achieved a significant reduction in HEVC encoding time at the cost of negligible increase in bit rate and negligible loss in PSNR. Please implement this. They suggest that their source code is an open source and can be used for research purposes only. (http://vision.poly.edu/~zma03/opensrc/sourceHM6.zip).

P.5.17 Please see P.5.16. The authors also suggest that similar approaches by other researchers (see Section 2 of this paper) can be combined with their work to further decrease the encoding time. See also [E43] and the references at the end of this paper. Explore this.

P.5.18 Please see P.5.17. The authors Zhang and Ma [E44, E149] also plan to explore the possibility of reducing the complexity of inter prediction modes. Investigate this.

P.5.19 Please see P.5.16 thru P.5.18. Combine both the complexity reduction techniques (intra/inter prediction modes) that can lead to practical HEVC encoders and evaluate the extent of complexity reduction in HEVC encoder with negligible loss in its compression performance.

Note that P.5.17 thru P.5.19 are research oriented projects leading to M.S. theses and Ph.D. dissertations.

P.5.20 Please access M. Zhang, C. Zhao and J. Xu, "An adaptive fast intra mode decision in HEVC", IEEE ICIP 2012, Orlando, FL, Sept.–Oct. 2012 [E43]. By utilizing the block's texture characteristics from rough mode decision and by further simplification of residual quad tree splitting process, their proposed method saves average encoding times 15% and 20% in the all intra high efficiency and all intra low complexity test conditions respectively with a marginal BD-rate increase [E81, E82, E96, E198]. Confirm these test results by implementing their approach.

P.5.21 See the paper by J. Nightingale, Q. Wang and C. Grecos, "HEVStream; A framework for streaming and evaluation of high efficiency video coding (HEVC) content in loss-prone networks', IEEE Trans. Consumer Electronics, vol. 59, pp. 404–412, May 2012 [E57]. They have designed and implemented a comprehensive streaming and evaluation framework for HEVC encoded video streams and tested its performance under a varied range of network conditions. Using some of the recommended test conditions (See Table III) the effects of applying bandwidth, packet loss, and path latency constraints on the quality (PSNR) of received video streams are reported. Implement and verify these tests. Besides PSNR, use SSIM [SSIM1] and BD rates [E81, E82, E96, E198] as benchmarks for comparison purposes.

P.5.22 See P.5.21. In terms of future work, the authors propose to focus on the development of suitable packet/NAL unit prioritization schemes for use in selective dropping schemes for HEVC. Explore this as further research followed by conclusions.

P.5.23 See the paper D. Marpe et al., "Improved video compression technology and the emerging high efficiency video coding standard", IEEE International Conf. on Consumer Electronics, pp. 52–56, Berlin, Germany, Sept. 2011 [E58]. The authors on behalf of Fraunhofer HHI have proposed a newly developed video coding scheme leading to about 30% bit rate savings compared to H.264/AVC HP (high profile) at the cost of significant increase in computational complexity. Several new features that contribute to the bit rate reduction have been explored. Implement this proposal and verify the bandwidth reduction. Explore the various techniques that were successfully used in reducing the complexity of H.264/AVC encoders. Hopefully these and other approaches can result in similar complexity reduction of HEVC encoders.

P.5.24 See the paper, M. Budagavi and V. Sze, "Unified forward + inverse transform architecture for HEVC', IEEE ICIP 2012, Orlando, FL, Sept.–Oct., 2012 [E35]. They take advantage of several symmetry properties of the HEVC core transform and show that the unified implementation (embedding multiple block size transforms, symmetry between forward and inverse transforms, etc.) results in 43–45% less area than separate forward and inverse core transform implementations. They show the unified forward + inverse 4-point and 8-point transform architectures in Figures 2 and 3 respectively. Develop similar architectures for the unified forward + inverse 16-point and 32-point transforms. Note that this requires developing equations for the 16 and 32 point transforms similar to those described in Equations 10–17 of this paper.

P.5.25 See P.5.24 The authors claim that the hardware sharing between forward and inverse transforms has enabled an area reduction of over 40%. Verify this.

P.5.26 In the transcoding arena, several researchers have developed, designed, tested and evaluated transcoders among H.264/AVC, AVS China, DIRAC, MPEG-2 and VC-1. Develop a transcoding system between H.264/AVC and HEVC (main profile) [E181]. Use HM9 [See E93].

P.5.27 Repeat P.5.26 for transcoding between MPEG-2 and HEVC (main profile), See [E98].

P.5.28 Repeat P.5.26 for transcoding between DIRAC and HEVC (main profile).

P.5.29 Repeat P.5.26 for transcoding between VC-1 and HEVC (main profile).

P.5.30 Repeat P.5.26 for transcoding between AVS China and HEVC (main profile).

P.5.31 As with H.264/AVC, HEVC covers only video coding. To be practical and useful for the consumer, audio needs to be integrated with HEVC encoded video. Encode HEVC video along with audio coder such as AAC, HEAAC, etc. following the multiplexing the coded bit streams at the transmitter. Demultiplexing the two bit streams, followed by decoding the audio and video while maintaining the lip sync is the role of the receiver. Implement these schemes for various video spatial resolutions and multiple channel audio. This comprises of several research areas at M.S. and doctoral levels. Such integrated schemes have been implemented for H.264/AVC, DIRAC and AVS China video with audio coders.

P.5.32 Similar to H.264/AVC for high video quality required within the broadcast studios (not for transmission/distribution), HEVC intra frame coding only can be explored. Compare this (HEVC intra-frame coding only) with H.264/AVC intra-frame coding only and JPEG 2000 at various bit rates using different test sequences. Use MSE/PSNR/SSIM/BD rates [E81, E82, E96, E198] and implementation complexity as comparison metrics.

P.5.33 In [E62], Ohm et al. compared the coding efficiency of HEVC at different bit rates using various test sequences with the earlier standards such as H.262/MPEG-2 video, H.263, MPEG-4 Visual (part 2) and H.264/AVC using PSNR and subjective quality as the metrics. They also indicate that software and test sequences for reproducing the selected results can be accessed from ftp://ftp.hhi.de/ieee-tcsvt/2012/

Repeat these tests and validate their results. Note that the DSIS used for measuring the subjective quality requires enormous test facilities, subjects (novices and experts) and may be beyond the availability of many research labs.

P.5.34 Repeat P.5.33 using SSIM [SSIM1] and BD-rates [E81, E82, E96, E198] as the performance metric and evaluate how these results compare with those based on PSNR.

P.5.35 Horowitz et al. [E66] compared the subjective quality (subjective viewing experiments carried out in double blind fashion) of HEVC (HM7.1) – main profile/low delay configuration – and H.264/AVC high profile (JM18.3) for low delay applications as in P.5.31 using various test sequences at different bit rates. To compliment these results, production quality H.264/AVC encoder known as x264 is compared with a production quality HEVC implementation from eBrisk Video (VideoLAN x264 software library, http://www.videolan.org/developers/x264.html version core 122 r2184, March 2012). They conclude that HEVC generally produced better subjective quality compared with H.264/AVC for low-delay applications at approximately 50% average bit rate of the latter. Note that the x264 configuration setting details are available from the authors on request. Several papers related to subjective quality/tests are cited in [E46]. Repeat these tests using PSNR,

BD rate [E81, E82, E96, E198] and SSIM [SSIM1] as the performance metrics and evaluate how these metrics can be related to the subjective quality.

P.5.36 Bossen et al. [E63] present a detailed and comprehensive coverage of HEVC complexity (both encoders and decoders) and compare with H.264/AVC high profile. They conclude for similar visual quality HEVC encoder is several times more complex than that of H.264/AVC. The payoff is HEVC accomplishes the same visual quality as that of H.264/AVC at half the bit rate required for H.264/AVC. The HEVC decoder complexity, on the other hand, is similar to that of H.264/AVC. They claim that hand held/mobile devices, lap tops, desk tops, tablets etc. can decode and display the encoded video bit stream. Thus real time HEVC decoders are practical and feasible. Their optimized software decoder (no claims are made as to its optimality) does not rely on multiple threads and without any parallelization using ARM and x64 computer. Implement this software for several test sequences at different bit rates and explore additional avenues for further optimization. See also [E192].

P.5.37 One of the three profiles in HEVC listed in FDIS (Jan. 2013) is intraframe (image) coding only. Implement this coding mode in HEVC and compare with other image coding standards such as JPEG, JPEG2000, JPEG-LS, JPEG-XR and JPEG using MSE/PSNR, SSIM [SSIM1] and BD-rate [E81, E82, E96, E198] as the metrics. As before, perform this comparison using various test sequences at different spatial resolutions and bit rates. See P.5.47.

P.5.38 Besides multiview/3D video [E34, E39], scalable video coding (temporal, spatial SNR-quality and hybrid) is one of the extensions/additions to HEVC [E325]. Scalable video coding (SVC) at present is limited to two layers (base layer and enhancement layer). SVC is one of the extensions in H.264/AVC and a special issue on this has been published [E69]. Software for SVC is available on line http://ip.hhi.de/omagecom_GI/savce/downloads/SVC-Reference-software.htm [E70]. Design, develop and implement these three different scalabilities in HEVC.

P.5.39 Sze and Budagavi [E67] have proposed several techniques in implementing CABAC (major challenge in HEVC) resulting in higher throughput, higher processing speed and reduced hardware cost without affecting the high coding efficiency. Review these techniques in detail and confirm these gains.

P.5.40 In [E73] details of the deblocking filter in HEVC are explained clearly. They show that this filter has lower computational complexity and better parallelization on multi cores besides significant reduction in visual artifacts compared to the deblocking filter in H.264/AVC. They validate these conclusions by using test sequences based on three configurations; 1) All-intra, 2) Random access and 3) Low delay. Go through this paper and related references cited at the end and confirm these results by running the simulations.
In [E209], deblocking filter is implemented in Verilog HDL.

P.5.41 Lakshman, Schwartz and Wiegand [E71] have developed a generalized interpolation framework using maximal-order interpolation with minimal support (MOMS) for estimating fractional pels in motion compensated prediction. Their technique shows improved performance compared to 6-tap and 12-tap filters [E109], specially for sequences with fine spatial details. This however may increase the complexity and latency. Develop parallel processing techniques to reduce the latency.

P.5.42 See P.5.41. Source code, complexity analysis and test results can be downloaded from
H. Lakshman et al., "CE3: Luma interpolation using MOMS", JCT-VC D056, Jan. 2011".
http://phenix.int-evry.fr/jct/doc_end_user/documents/4_Deagu/wg11/JCTVC-D056-v2.zip. See [E71]. Carry out this complexity analysis in detail.

P.5.43 Correa et al. [E84] have investigated the coding efficiency and computational complexity of HEVC encoders. Implement this analysis by considering the set of 16 different encoding configurations.

P.5.44 See P.5.43 Show that the low complexity encoding configurations achieve coding efficiency comparable to that of high complexity encoders as described in draft 8 [E60].

P.5.45 See P.5.43 Efficiency and complexity analysis explored by Correa et al. [E84] included the tools (non-square transform, adaptive loop filter and LM luma) which have been subsequently removed in the HEVC draft standard [E60]. Carry out this analysis by dropping these three tools.

P.5.46 Schierl et al. in their paper "System layer integration of HEVC" [E83] suggest that the use of error concealment in HEVC should be carefully considered in implementation and is a topic for further research. Go through this paper in detail and explore various error resilience tools in HEVC. Please note that many error resilience tools of H.264/AVC such as FMO, ASO, redundant slices, data partitioning and SP/SI pictures have been removed due to their very rare deployment in real-world applications.

P.5.47 Implement the lossless coding of HEVC main profile (Figure 5.13) proposed by Zhou et al. [E85] and validate their results. Also compare with current lossless coding methods such as JPEG-2000, etc., based on several test sequences at different resolutions and bit rates. Comparison metrics are PSNR/MSE, SSIM, BD-rates [E81, E82, E96, E198], etc. Consider the implementation complexity also in the comparison.

Cai et al. [E112] have also compared the performance of HEVC, H.264/AVC, JPEG2000 and JPEG-LS for both lossy and lossless modes. For lossy mode their comparison is based on $\text{PSNR}_{\text{avg}} = (6 \times \text{PSNR}_y + \text{PSNR}_u + \text{PSNR}_v)/8$ only. This is for 4:2:0 format and is by default. Extend this comparison based on SSIM, BD-rate [81, E82, E96, E198] and implementation complexity. Include also JPEG-XR which is based on HD-Photo of Microsoft in this comparison. They have provided an extensive list of references related to performance comparison of intra coding of several standards. See also P.5.37. See also [E145].

P.5.48 See [E95]. An efficient transcoder for H.264/AVC to HEVC by using a modified MV reuse has been developed. This also includes complexity scalability trading off RD performance for complexity reduction. Implement this. Access references [4–7] related to transcoding overview papers cited at the end of [E95]. See also [E98], [E146], [E148] and [E333].

P.5.49 See P.5.48. The authors in [E95] suggest that more of the H.264/AVC information can be reused in the transcoder to further reduce the transcoder complexity as future work. Explore this in detail and see how the transcoder complexity can be further reduced. The developed techniques must be justified based on the comparison metrics (See P.5.47).

P.5.50 See P.5.48 and P.5.49. Several other transcoders can be developed. i.e.,

A) Transcoder between MPEG-2 and HEVC (there are still many decoders based on MPEG-2). Please access [E98] T. Shanableh, E. Peixoto and E. Izquierdo, "MPEG-2 to HEVC video transcoding with content-based modeling", IEEE Trans. CSVT, vol. 23, pp. 1191–1196, July 2013. The authors have developed an efficient transcoder based on content-based machine learning. In the conclusions section, they have proposed future work. Explore this. In the abstract they state "Since this is the first work to report on MPEG-2 to HEVC video transcoding, the reported results can be used as a benchmark for future transcoding research". This is a challenging research in the transcoding arena.
B) Transcoder between AVS China and HEVC.
C) Transcoder between VC-1 and HEVC.
D) Transcoder between VP9 and HEVC.

Implement these transcoders. Note that these research projects are at the M.S. thesis levels.
You can access the theses related to transcoders that have been implemented as M.S. theses from the web site http://www.uta.edu/faculty/krrao/dip, click on courses and then click on EE5359. Or access directly http://www.uta.edu/faculty/krrao/dip/Courses/EE5359/index.html

P.5.51 Please access [E72]. This paper describes low complexity-high performance video coding proposed to HEVC standardization effort during its early stages of development. Parts of this proposal have been adopted into TMuC. This proposal is called Tandberg, Ericsson and Nokia test model (TENTM). Implement this proposal and validate the results. TENTM proposal can be accessed from reference 5 cited at the end of this paper.

P.5.52 Reference 3 (also web site) cited in [E72] refers to video coding technology proposal by Samsung and BBC [Online]. Implement this proposal.

P.5.53 Reference 4 (also web site) cited in [E72] refers to video coding technology proposal by Fraunhoff HHI [Online]. Implement this proposal.

P.5.54 Please access [E106] M.S. Thesis by S. Gangavathi entitled, "Complexity reduction of H.264 using parallel programming"

M.S. Thesis, EE Dept., University of Texas at Arlington, Arlington, Texas, Dec. 2012. http://www.uta.edu/faculty/krrao/dip click on courses and then click on EE5359 Scroll down and go to Thesis/Project Title and click on S. Gangavathi.
By using CUDA he has reduced the H.264 encoder complexity by 50% in the baseline profile. Extend this to Main and High profiles of H.264.

P.5.55 See P.5.54. Extend Gangavathi's approach to HEVC using several test sequences coded at different bit rates. Show the performance results in terms of encoder complexity reduction and evaluate this approach based on SSIM [SSIM1], BD-PSNR, BD- bit rates [E81, E82, E96, E198] and PSNR as the metrics.
UTA/EE5359 course web site: http://www-ee.uta.edu/Dip/Courses/EE5359/index.html

P.5.56 Zhang, Li and Li [E108] have developed a gradient-based fast decision algorithm for intra prediction in HEVC. This includes both prediction unit (PU) size and angular prediction modes. They claim a 56.7% savings of the encoding time in intra HE setting and up to 70.86% in intra low-complexity setting compared to the HM software [E97]. Implement this and validate their results.

P.5.57 Please see P.5.56 In the section Conclusion the authors suggest future work on how to obtain the precise coding unit partition for the complex texture picture combined with RDO technique used in HEVC. Explore this.

P.5.58 Wang et al. [E110] present a study of multiple sign bit hiding scheme adopted in HEVC. This technique addresses the joint design of quantization transform coefficient coding using the data hiding approach. They also show that this method consistently improves the rate-distortion performance for all standard test images resulting in overall coding gain in HEVC. In terms of future work, they suggest that additional gains can be expected by applying the data hiding technique to other syntax elements. Explore this.

P.5.59 Please see P.5.58. The authors comment that the general problem of joint quantization and entropy coding design remains open. Explore this.

P.5.60 Lv et al. [E116] have developed a fast and efficient method for accelerating the quarter-pel interpolation for ME/MC using SIMD instructions on ARM processor. They claim that this is five times faster than that based on the HEVC reference software HM 5.2. See section V acceleration results for details. Using NEON technology verify their results.

P.5.61 Shi, Sun and Wu [E76] have developed an efficient spatially scalable video coding (SSVC) for HEVC. Using two layer inter prediction schemes. Using some test sequences they demonstrate the superiority of their technique compared with other SSVC schemes. Implement this and validate their results. In the conclusion section, they suggest future work to further improve the performance of their scheme. Explore this in detail. Review the papers related to SVC listed at the end in references. See H. Schwarz, D. Marpe and T. Wiegand, "Overview of the Scalable Video Coding Extension of the H.264/AVC Standard", *IEEE Trans. on Circuits and Systems for Video Technology,* vol. 17, pp. 1103–1120, Sept. 2007. This is a special issue on SVC. There are many other papers on SVC.

P.5.62 Zhou et al. [E85, E123] have implemented HEVC lossless coding for main profile by simply bypassing transform, quantization and in-loop filters and compared with other lossless coding methods such as JPEG-2000, ZIP, 7-Zip, WinRAR etc. Implement this and also compare with JPEG, JPEG-XR, PNG, etc. Consider implementation complexity as another metric.

P.5.63 References [E126–E129, E158–E160, E325] among others address scalable video coding extensions to HEVC. Review these and implement spatial/quality (SNR)/temporal scalabilities. See also P.5.61.

P.5.64 Please access [E66]. In this paper Horowitz et al. demonstrate that HEVC yields similar subjective quality at half the bit rate of H.264/AVC using both HM 7.1 and JM 18.3 softwares. Similar conclusions are also made using eBrisk and x264 softwares. Using the latest HM software, conduct similar tests on the video test sequences and confirm these results. Consider implementation complexity as another comparison metric.

P.5.65 Kim et al. [E134] developed a fast intra-mode (Figures 5.8 and 5.9) decision based on the difference between the minimum and second minimum SATD-based (sum of absolute transform differences) RD cost estimation and a fast CU-size (Figure 5.7) decision based on RD cost of the best intra mode. Other details are described in this paper. Based on the simulations conducted on class A and class B test sequences (Table 5.1), they claim that their proposed method achieves an average time reduction (ATR) of 49.04% in luma intra-prediction and an ATR of 32.74% in total encoding time compared to the HM 2.1 encoder. Implement their method and confirm these results. Use the latest HM software [E56]. Extend the luma intra-prediction to chroma components also.

P.5.66 Flynn, Martyin-Cocher and He [E135] proposed to JCT-VC best effort decoding a 10-bit video bit stream using an 8-bit decoder. Simulations using several test sequences based on two techniques a) 8-bit decoding by adjusting inverse transform scaling and b) hybrid 8-bit-10-bit decoding using rounding in picture construction process were carried out. HM-10 low-B main-10 sequences (F. Bossen, "Common HM test conditions and software reference configurations", JCT-VC-L1100, JCT-VC, Jan. 2013, E178) were decoded and the PSNR measured against the original input sequences. PSNR losses averaged 6 dB and 2.5 dB respectively for the two techniques compared against the PSNR of the normal (10 bit) decoder. Implement these techniques and confirm their results. Explore other techniques that can reduce the PSNR loss.

P.5.67 Pan et al. [E140] have developed two early terminations for TZSearch algorithm in HEVC motion estimation and have shown that these early terminations can achieve almost 39% encoding saving time with negligible loss in RD performance using several test sequences. Several references related to early terminations are listed at the end of this paper. Review these references and simulate the techniques proposed by Pan et al. and validate their conclusions. Extend these simulations using HDTV and ultra HDTV test sequences.

P.5.68 The joint call for proposals for scalable video coding extension of HEVC was issued in July 2012 and the standardization started in Oct. 2012 (See E76 and E126 thru E129, E158 thru E160, E325). Some details on scalable video codec design based on multi-loop and single-loop architectures are provided in [E141]. The authors in [E141] have developed a multi-loop scalable video coder for HEVC that provides a good complexity and coding efficiency trade off. Review this paper and simulate the multi-loop scalable video codec. Can any further design improvements be done on their codec? Please access the web sites below.

JSVM-joint scalable video model reference software for scalable video coding. (on line) http://ip.hhi.de/imagecom_GI/savce/downloads/SVC-Reference-software.htm
JSVM9 Joint scalable video model 9 http://ip.hhi.de/imagecom_GI/savce/dowloads
See H. Schwarz, D. Marpe and T. Wiegand, "Overview of the Scalable Video Coding Extension of the H.264/AVC Standard", *IEEE Trans. on Circuits and Systems for Video Technology*, vol. 17, pp. 1103–1120, Sept. 2007. This is a special issue on SVC. There are many other papers on SVC.

P.5.69 Tohidpour, Pourazad and Nasiopoulos [E142] proposed an early-termination interlayer motion prediction mode search in HEVC/SVC (quality/fidelity scalability) and demonstrate complexity reduction up to 85.77% with at most 3.51% bit rate increase (almost same PSNR). Simulate this approach. Can this be combined with the technique developed in [E141]—See P.5.68.

P.5.70 In [E143], Tan, Yeo and Li proposed a new lossless coding scheme for HEVC and also suggest how it can be incorporated into the HEVC coding framework. In [E85], Zhou et al. have implemented a HEVC lossless coding scheme. Compare these two approaches and evaluate their performances in terms of complexity, SSIM, BD-rates [E81, E82, E96, E198] and PSNR for different video sequences at various bit rates. See also P.5.62.

P.5.71 See the paper Y. Pan, D. Zhou and S. Goto, "An FPGA-based 4K UHDTV H.264/AVC decoder", IEEE ICME, San Jose, CA, July 2013. Can a similar HEVC decoder be developed? (Sony already has a 4K UHDTV receiver on the market).

P.5.72 In [E13], Asai et al. proposed a new video coding scheme optimized for high resolution video sources. This scheme follows the traditional block based MC+DCT hybrid coding approach similar to H.264/AVC, HEVC and other ITU-T/VCEG and ISO/IEC MPEG standards. By introducing various technical optimizations in each functional block, they achieve roughly 26% bit-rate savings on average compared to H.264/AVC high profile at the same PSNR. They also suggest improved measures for complexity comparison. Go through this paper in detail and implement the codec. Consider BD-PSNR and BD-Rate [E81, E82, E96, E198] and SSIM [SSIM1] as the comparison metrics. For evaluating the implementation complexity, consider both encoders and decoders. Can this video coding scheme be further improved? Explore all options.

P.5.73 See P.5.72 By bypassing some functional blocks such as transform/quantization/in loop deblocking filter, Zhou et al. [E85, E123] have implemented a HEVC lossless coding scheme (Figure 5.13). Can a similar lossless coding scheme be implemented in the codec proposed by Asai et al. [E13]. If so compare its performance with other lossless coding schemes. See P.5.62.

P.5.74 In [E145], the authors propose HEVC intra coding acceleration based on tree level mode correlation. They claim a reduction by up to 37.85% in encoding processing time compared with HM4.0 intra prediction algorithm with negligible BD-PSNR loss [E81, E82, E96, E198]. Implement their algorithm and confirm the results using various test sequences. Use the latest HM instead of HM 4.0.

P.5.75 See P.5.48 In [E146], Peixoto et al. have developed a H.264/AVC to HEVC video transcoder which uses machine learning techniques to map H.264/AVC macroblocks into HEVC coding units. Implement this transcoder. In the conclusions, the authors suggest ways and methods by which the transcoder performance can be improved (future work). Explore these options. See [E263], where different methods such as mode mapping, machine learning, complexity scalable and background modeling are discussed and applied to H.264/AVC to HEVC transcoder. (Also see [E267])

P.5.76 Lainema et al. [E78] give a detailed description of intra coding of the HEVC standard developed by the JCT-VC. Based on different test sequences, they demonstrate significant improvements in both subjective and objective metrics over H.264/AVC and also carry out a complexity analysis of the decoder. In the conclusions, they state "Potential future work in the area includes e.g., extending and tuning the tools for multiview/scalable coding, higher dynamic range operation and 4:4:4 sampling formats". Investigate this future work thoroughly.

P.5.77 Pl access the paper [E175], Y. Tew and K. S. Wong, "An overview of information hiding in H.264/AVC compressed video", IEEE Trans. CSVT, vol. 24, pp. 305–319, Feb. 2014. This is an excellent review paper on information hiding specially in H.264/AVC compressed domain. Consider implementing information hiding in HEVC (H.265) compressed domain.

Abstract is reproduced below:
Abstract—Information hiding refers to the process of inserting information into a host to serve specific purpose(s). In this article, information hiding methods in the H.264/AVC compressed video domain are surveyed. First, the general framework of information hiding is conceptualized by relating state of an entity to a meaning (i.e., sequences of bits). This concept is illustrated by using various data representation schemes such as bit plane replacement, spread spectrum, histogram manipulation, divisibility, mapping rules and matrix encoding. Venues at which information hiding takes place are then identified, including prediction process, transformation, quantization and entropy coding. Related information hiding methods at each venue are briefly reviewed, along with the presentation of the targeted applications, appropriate diagrams and references. A timeline diagram is constructed to chronologically summarize the invention of information hiding methods in the compressed still image and video domains since year 1992. Comparison among the considered information hiding methods is also conducted in terms of venue, payload, bitstream size overhead, video quality, computational complexity and video criteria. Further perspectives and recommendations are presented to provide a better understanding on the current trend of information hiding and to identify new opportunities for information hiding in compressed video.
Implement similar approaches in HEVC – various profiles. The authors suggest this, among others, as future work in VII Recommendation and further research direction and VIII Conclusion sections. These two sections can lead to several projects/theses.

P.5.78 Pl access X.-F. Wang and D.-B. Zhao, “Performance comparison of AVS and H.264/AVC video coding standards”, J. Comput. Sci. & Tech., vol. 21, pp. 310–314, May 2006. Implement similar performance comparison between AVS China and HEVC (various profiles).

P.5.79 See [E152] about combining template matching prediction and block motion compensation. Go through several papers on template matching listed in the references at the end of this paper and investigate/evaluate thoroughly effects of template matching in video coding.

P.5.80 See P.5.79 In [E152], the authors have developed the inter frame prediction technique combining template matching prediction and block motion compensation for high efficiency video coding. This technique results in 1.7–2.0% BD-rate [E81, E82, E96, E198] reduction at a cost of 26% and 39% increase in encoding and decoding times respectively based on HM-6.0. Confirm these results using the latest HM software. In the conclusions the authors state "These open issues need further investigation". Explore these.

P.5.81 In [E120] detailed analysis of decoder side motion vector derivation (DMVD) and its inclusion in call for proposals for HEVC is thoroughly presented. See also S. Kamp, *Decoder-Side Motion Vector Derivation for Hybrid Video Coding*, (Aachen Series on Multimedia and Communications Engineering Series). Aachen, Germany: Shaker-Verlag, Dec. 2011, no. 9. The DMVD results in very moderate bit rate reduction, however, offset by increase in decoder side computational resources. Investigate this thoroughly and confirm the conclusions in [E120].

P.5.82 Implement, evaluate and compare Daala codec with HEVC. Daala is the collaboration between Mozilla foundation and Xiph.org foundation. It is an open source codec. Access details on Daala codec from Google. "The goal of the DAALA project is to provide a free to implement, use and distribute digital media format and reference implementation with technical performance superior to H.265".

P.5.83 Pl access the thesis 'Complexity Reduction for HEVC intraframe Luma mode decision using image statistics and neutral networks', by D.P. Kumar from UTA -MPL web site EE5359. Extend this approach to HEVC interframe coding using neural networks. Kumar has kindly agreed to help any way he can on this. This extension is actually a M.S. thesis.

P.5.84 Pl see P.5.83. Combine both interframe and intraframe HEVC coding to achieve complexity reduction using neural networks.

P.5.85 HEVC range extensions include screen content (text, graphics, icons, logos, lines etc.) coding [E322]. Combination of natural video and screen content has gained importance in view of applications such as wireless displays, automotive infotainment, remote desktop, distance education, cloud computing, video walls in control rooms etc. These papers specifically develop techniques that address screen content coding within the HEVC framework. Review and implement these techniques and also explore future work. IEEE Journal on Emerging and Selected Topics in Circuits and Systems (JETCAS) has called for papers for the special Issue on Screen Content Video

Coding and Applications. Many topics related to screen content coding (SCC) and display stream compression (see P.5.85a) are suggested. These can lead to several projects and theses. This special issue was published in Dec. 2016 [SCC15].

P.5.85a The Video Electronics Standards Association also recently completed a Display Stream Compression (DSC) standard for next-generation mobile or TV/Computer display interfaces. The development of these standards introduced many new ideas, which are expected to inspire more future innovations and benefit the varied usage of screen content coding. Review this standard in detail and implement DSC for next-generation mobile or TV/Computer display interfaces.

P.5.86 Access the M.S Thesis (2011) by P. Ramolia, "Low complexity AVS-M using machine learning algorithm C4.5", from UTA web site. Explore and implement similar complexity reduction in HEVC – all profiles – using machine learning.

P.5.87 The papers cited in [E161–E163] deal with DSP based implementation of HEVC9.0 decoder. They state that with an optimization process HEVC decoders for SD resolution can be achieved with a single DSP. Implement this.

P.5.88 See P.5.87. The authors state that for HD formats multi-DSP technology is needed. Explore this.

P.5.89 In [E168], the authors state that the motion estimation (ME) for ultra-high definition (UHD) HEVC requires 77–81% of computation time in the encoder. They developed a fast and low complexity ME algorithm and architecture which uses less than 1% of the number of operations compared to full search algorithm at negligible loss in BD bit rate [E81, E82, E96, E198]. They claim that the proposed ME algorithm has the smallest hardware size and the lowest power consumption among existing ME engines for HEVC. Review this paper along with the references (listed at the end) related to ME algorithms. Implement their algorithm and verify the results and conclusions based on various UHD test sequences. Consider also BD-PSNR as another performance metric. Review the paper, M. Jakubowski and G. Pastuzak, "Block based motion estimation algorithms – a survey", Opto-Electronics review, vol. 21, no. 1, pp. 86–102, 2013.

P.5.90 See P.5.89 Implement the fast and low complexity ME algorithm and architecture in HEVC encoder using super hi-vision test sequences. See [E164].

P.5.91 In Section 5.10 Summary some references related to reducing the implementation complexity of HEVC encoder (includes both intra and inter predictions) are cited. Review these and other related papers and develop a technique that can significantly reduce HEVC encoder complexity with negligible loss in BD-PSNR and BD bit rates [E81, E82, E96, E198]. Demonstrate this based on various test sequences at different bit rates and resolutions.

P.5.92 Fast early zero block detection is developed for HEVC resulting in substantial savings in encoder implementation. Go through [E166, E171] and related papers and implement this technique in HEVC encoder for all profiles and levels. Compare the complexity reduction with HM software using BD bit rates, BDPSNR, [E81, E82, E96, E198]. SSIM [SSIM1] and computational time as performance metrics.

P.5.93 Nguyen et al. [E173] present a detailed overview of transform coding techniques in HEVC focusing on residual quadtree structure and entropy coding structure. Using the standard test sequences and test conditions they show the resulting improved coding efficiency. Implement their techniques and confirm the results (Tables I–VI). Use the latest HM. If possible try also 4k × 2k test sequences. This project is highly complex and can be subdivided into various sections.

P.5.94 See P.5.85. Naccari et al. [E157] have specifically developed residual DPCM (RDPCM) for lossless screen content coding in HEVC and demonstrated that their algorithm achieves up to 8% average bit rate reduction with negligible increase in decoder complexity using the screen content coding set. Confirm their findings by implementing the RDPCM.

P.5.95 See P.5.94 The authors state that future work involves extension of the inter RDPCM tool for the lossy coding at very high bit rates leading to visually lossless coding. Explore this in detail.

P.5.96 Lv et al. [E109] carried out a detailed analysis of fractional pel interpolation filters both in H.264/AVC and in HEVC. They also conducted complexity analysis of these filters. By comparing the contribution of these filters to the compression performance they conclude that the interpolation filters contribute 10% performance gain to HEVC [E45] compared to H.264/AVC at the cost of increased complexity. They also demonstrate that the frequency responses of quarter pel filters in HEVC are superior to those in H.264/AVC. Their conclusions are based on reference software HM 5.2 using low resolution (416 × 240) test sequences. Carry out this analysis using HM 15.0 software and high resolution test sequences.

P.5.97 See P.5.96. Go through the analysis of how DCTIF (discrete cosine transform interpolation filters) have been developed [E109]. By plotting the frequency responses of the filters in H.264/AVC and in HEVC [E45], they show that in the passband the filters in HEVC are much flatter and have much smaller ripples than those in H.264/AVC (Figure 5 in [E107]). Confirm these conclusions.

P.5.98 Na and Kim [E176] developed a detailed analysis of no-reference PSNR estimation affected by deblocking filtering in H.264/AVC bit streams. Go through this paper in detail and develop similar analysis for HEVC. Note that in HEVC there are two in loop filters, deblocking, and SAO filters.

P.5.99 Grois et al. [E177] conducted a detailed performance comparison of H.265/MPEG-HEVC, VP9 and H.264/MPEG-AVC encoders. Using various test sequences and common HM test conditions [E176]. They conclude that H.265/MPEG-HEVC provides average bit-rate savings of 43.3 and 39.35% relative to VP9 and H.264/MPEG-AVC encoders, respectively. However H.265/MPEG-HEVC reference software requires 7.35 times that of VP9 encoding time. On the other hand VP9 requires 132.6 times that of x264 encoding time. These results are based on equal $\mathrm{PSNR}_{\mathrm{YUV}}$. Implement these performance comparison tests using the HM15.0 software. See the references cited at the end of [E177]. Please see P.5.33, P.5.36, and P.5.64.

P.5.100 See P.5.99. Review the Tables IV and VI in [E 177]. Develop bit-rate savings between VP9 and ×264 (Table IV) for all the test sequences. Develop encoding run times between H.265/MPEG-HEVC and x264 (Table VI) for all the test sequences. Both these parameters need to be based on equal PSNR_{YUV}.

P.5.101 Abdelazim, Masri, and Noaman [E179] have developed ME optimization tools for HEVC. With these tools they are able to reduce the encoder complexity up to 18% compared to the HM software at the same PSNR and bit rate using various test sequences. Implement their scheme and confirm the results.

P.5.102 Ohm et al. [E62] have compared the HEVC coding efficiency with other video coding standards such as H.264/MPEG-4 AVC, H.264/MPEG-4 visual (part 2), H.262/MPEG-2 part 2 video [E183] and H.263. This comparison is based on several standard test sequences at various bit rates. The comparison metrics include PSNR and mean opinion score (MOS) [E180] based on double stimulus impairment scale (DSIS) both for interactive and entertainment applications. The comparison, however, does not include implementation complexity. Implement this metric for all the video coding standards and draw the conclusions.

P.5.103 See P.5.102 Extend the coding efficiency comparison between HEVC and DIRAC. Software and data for reproducing selected results can be found at ftp://ftp.hhi.de/ieee-tcsvt/2012

P.5.104 See P.5.102 Extend the coding efficiency comparison between HEVC and VC-1.

P.5.105 See P.5.102 Extend the coding efficiency comparison between HEVC and AVS China.

P.5.106 See P.5.102 Extend the coding efficiency comparison between HEVC and VP9.

P.5.107 See [E190] Rerabek and Ebrahimi have compared the compression efficiency between HEVC and VP9 based on image quality. Implement this comparison based on BD-PSNR, BD-bit rate [E81, E82, E96, E198], implementation complexity etc using the UHDTV sequences.

P.5.108 See P.5.107. The authors [E190], in the conclusion, suggest extending this comparison to Internet streaming scenarios. Implement this.

P.5.109 Grois et al. [E190] have made a comparative assessment of HEVC/H.264/VP9 encoders for low delay applications using 1280×720 60 fps test sequences. Implement this. Extend this comparison to UHDTV sequences. Consider implementation complexity as another metric.

P.5.110 Heindel, Wige and Kaup [E193] have proposed lossy to lossless scalable video coding. This is based on lossy base layer (BL) HEVC encoder and lossless enhancement layer (EL) encoder. They claim that this approach has advantages over the scalable extension of the HEVC. Implement this technique and confirm their results. They have suggested future work. Explore this.

P.5.111 See P.5.110. Heindel, Wige and Kaup [E193] have investigated different prediction schemes for the EL beyond [E192] and compared the performance with single layer JPEG-LS (lossless compression) (See references 8 and 9 in [E193]). Implement this.

P.5.112 Bross et al. [E192] have demonstrated that real time HEVC decoding of 4K (3840 × 2160) video sequences is feasible in current desktop CPUs using four CPU cores. Implement this using the EBU HD-1 TEST SET (available http://tech.ebu.ch/testsequences/uhd-1).

P.5.113, Vanne, Viitanen and Hamalainen [E195] have optimized inter prediction mode decision schemes for HEVC encoder and have reduced the encoder complexity (compared to HM 8.0) by 31%–51% at a cost of 0.2%–1.3% for random access (similar ranges for low delay). They have used the test sequences (Table VI) specified in F. Bossen, "Common test conditions and software reference configurations", Document JCTVC-J1100, Incheon, S. Korea, April 2013. Their mode decision schemes are described in Section V – proposed mode decision schemes. Implement their complexity reduction techniques in the HEVC encoder.

P.5.113 See P.5.112 Extend this to ultra HDTV (4K resolution) and to super hi-vision (8K resolution) [E164] test sequences.

P.5.114 See P.5.113. In section VII Conclusion the authors propose to combine the proposed techniques with the other existing approaches (see the references at the end) without compromising the RD performance or losing the ability for parallelization and hardware acceleration. Explore this.

P.5.115 Several lossless image compression techniques have been developed-HEVC, H.264, JPEG, JPEG-LS, JPEG 2000, 7-Zip, WinRAR etc. [E85, E114, E147, E194, E196]. Compare these techniques in terms of implementation complexity using several test images at different resolutions.

P.5.116 Several techniques [E166, E171, E186] for detection of zero blocks in HEVC have been investigated with a view to reduce the encoder complexity. Review these in detail and also the techniques developed for H.264 (See the references at the end of these papers) and develop an optimal approach for zero block detection in HEVC. The optimal approach needs to be justified in terms of encoder complexity and comparison metrics such as BD-PSNR, BD-bitrate [E81, E82, E96, E198] and visual quality using several test sequences.

P.5.117 Chapter 6, Budagavi, Fuldseth and Bjontegaard, "HEVC transform and quantization", in Tabatabai et al. [E202] describes the 4×4, 8×8, 16×16 and 32×32 integer DCTs including the embedding process (small size transforms are embedded in large size transforms) Review the references listed at the end of this chapter. The inverse integer DCTs are transposes of the corresponding integer DCTs. Compute the orthogonal property of these integer DCTs. Hint; Matrix multiply the integer DCTs with their corresponding transposes.
See J. Sole et al, "Transform coefficient coding in HEVC", IEEE Trans. CSVT, Vol. 22, pp. 1765–1777, Dec. 2012. See also M. Budagavi et al., "Core transform design in the high efficient video coding standard (HEVC)", IEEE J. of selected topics in signal processing, vol. 7, pp. 1029–1041, Dec. 2013. The philosophy in designing these integer DCTs (core transform) specially meeting some properties can be observed in this chapter. Review this and justify why alternative proposals to the core transform design were not selected by the JCT-VC.

P.5.118 A fast inter-mode decision algorithm for HEVC is proposed by Shen, Zhang and Liu [E199]. This algorithm is based on jointly using the inter-level correlation of quad-tree structure and spatiotemporal correlation. Using several test sequences, they show that the proposed algorithm can save 49% to 52%computational complexity on average with negligible loss of coding efficiency (see Tables X thru XIII and Figures 5 thru 8). They also show the superiority of their algorithm compared to other algorithms (see references at the end) in terms of coding time saving (Figure 9a and Figure 10a) with negligible BDBR increase (see Figure 9b and Figure 10b). Implement this algorithm and validate their results. Explore this algorithm for further improvements in reducing the encoder complexity.

P.5.119 Peng and Cosman [E200] have developed a weighted boundary matching error concealment (EC) technique for HEVC and have demonstrated its superiority over other EC methods (see references 4–6 cited at the end). For simulation two video sequences BQMall and Drill (832×480) at 50 fps are used. The results are based on QP 28 using HM 11. Consider various other test sequences including higher resolutions (HDTV and ultra HDTV), using more QPs and HM15. Show the results in terms of PSNR versus error rate

(Figure 3) and the frames (original, loss pattern and error concealed (Figure 4)).

P.5.120 Warrier [E203] has implemented encode/decode, multiplex/demultiplex HEVC video/AAC audio while maintaining lip synch. She used main profile/intra prediction in HEVC. Extend this to inter prediction and also to other profiles.

P.5.121 Y. Umezaki and S. Goto [E205] have developed two techniques – partial decoding and tiled encoding – for region of interest (ROI) streaming in HEVC. They have compared these two methods using HM10.0 in terms of decoding cost, bandwidth efficiency and video quality. Regarding future work, they suggest combining tiled encoding and partial decoding to further reduce the decoding cost. Explore this future work and evaluate the reduction in the decoding cost. Use the latest HEVC reference software HM 15.0.

P.5.122 Mehta [E204] has effectively introduced parallel optimization of mode decision for intra prediction in HEVC and was able to reduce the encoding time by 35%–40% on average with negligible loss in image quality. In future work, he has suggested that there are many other effective techniques to implement parallelism to different sections of the HM software leading to further reduction in encoding time. Review the thesis and future work in detail and implement these techniques.

P.5.123 See [E206] and [E207} Scalable HEVC (SHVC) test model 6 is described in [E207]. Overview of the scalable extensions of the H.265/HEVC video coding standard is described in [E207]. Implement the SHVC using test sequences for scalabilities; Temporal, SNR, bit-depth, spatial, interlaced-to-progressive, color gamut, hybrid and their combinations. Note that each scalability, by itself, is a project. Compare the enhancement layer output with direct encoder/decoder (no scalability) output based on the standard metrics. See the papers below:

H. Schwarz, D. Marpe, and T. Wiegand, "Overview of Scalable Video Coding Extension of the H.264/AVC Standard", IEEE Trans. CSVT, vol. 17, pp. 1103–1120, Sept. 2007.

H. Schwarz and T. Wiegand, "The Scalable Video Coding Amendment of the H.264/AVC Standard", csdn.net, world pharos, blog. [online]. Available: http://blog.csdn.net/worldpharos/article/details/3369933 (accessed on June 20th 2014).

Also access Karuna Gubbi S.S. Sastri, "Efficient intra-mode decision for spatial scalable extension of HEVC", M.S. Thesis, EE Dept., University of Texas at Arlington, Arlington, Texas, Aug. 2014. http://www.uta.edu/faculty/krrao/dip click on courses and then click on EE5359 Scroll down and go to Thesis/Project Title and click on Karuna Gubbi S.S. Sastri

P.5.124 Whereas Mehta [E204] was able to reduce the encoding time in intra frame coding, Dubhashi [E204a] was able to reduce the time taken for the motion estimation process in inter frame coding in HEVC. Combine these two techniques to reduce the encoding time taken by the HEVC encoder. Consider various test sequences at different spatial and temporal resolutions.

P.5.125 Projects related to ATSC3.0 (Advanced Television Systems Committee) www.atsc.org

ATSC (Established in 1982) is a consortium of broadcasters, vendors and trade groups. ATSC 3.0
The Advanced Television Systems Committee is an international, non-profit organization developing voluntary standards for digital television. The ATSC member organizations represent the broadcast, broadcast equipment, motion picture, consumer electronics, computer, cable, satellite, and semiconductor industries. For more information visit www.atsc.org
Overall intent is to enable seamless transmission of HD, 4K video, 22.2 kHz audio and other data streams to fixed, mobile and handheld devices in all types of terrains. Industry, research institutes and academia have submitted proposals and are submitting proposals for different layers. Hence this is a fertile ground for R & D. Access these proposals and explore possible research topics. (See M.S. Richter et al., "The ATSC digital television system", Proc. IEEE, vol. 94, pp. 37–43, Jan. 2006. Special issue on global digital television technology, Proc. IEEE, vol. 94, Jan. 2006. In future this may be extended to 8K video.
See J.M. Boyce, "The U.S. digital television broadcasting transition", IEEE SP Magazine, vol. 29, pp. 108–112, May 2012.
ATSC approves mobile & handheld 20 July 2009.
Candidate Standard
ATSC DTV Moves into High Gear with Mobile and Handheld Specifications. The Advanced Television Systems Committee, Inc. (ATSC) has elevated its specification for Mobile Digital Television to Candidate Standard status. The new Mobile DTV Candidate Standard provides the technical capabilities

necessary for broadcasters' to provide new services to mobile and hand-held devices using their digital television transmissions. ATSC Mobile DTV includes a highly robust transmission system based on vestigial sideband (VSB) modulation coupled with a flexible and extensible IP based transport, efficient MPEG AVC (H.264) video and HE AAC v2 audio (ISO/IEC 14496-3) coding. (HE-AAC High efficiency advanced audio coder).
It seems ATSC is exploring HEVC for these services. This can lead to number of R&D projects.

P.5.126 It is stated that ATSC also is conducting using reference software and experimental evaluation methodology for the 3D-TV terrestrial broadcasting. Review [E211] and implement the real-time delivery of 3D-TV terrestrial broadcasting. Access also www.atsc.org

Please do extensive literature survey on ATSC before writing the thesis proposal.
Pl access documents related to ATSC 3.0:
[1] Report on ATSC 3.0:
Link: http://www.atsc.org/cms/pdf/pt2/PT2-046r11-Final-Report-on-NGBT.pdf
[2] Presentation slides on ATSC 3.0:
Link: https://mentor.ieee.org/802.18/dcn/12/18-12-0011-00-0000-nab-presentation-on-atsc-3.pdf
[3] List of ATSC standards and an overview of ATSC:
Link: http://www.atsc.org/cms/pdf/ATSC2013_PDF.pdf
[4] Link to ATSC standards
Link: http://www.atsc.org/cms/index.php/standards/standards?layout=default?

P.5.128 Wang, Zhou and Goto [E212] have proposed a motion compensation architecture that can support real-time decoding 7680×4320@30fps at 280 MHz. Develop a similar architecture for ME/MC that can support real time encoder for ultra HDTV. See also M. Tikekar et al., "Decoder architecture for HEVC," Chapter 10 in Tabatabai et al. [E202].

P.5.129 VESA/DSC Pl see below:
The Video Electronics Standards Association (VESA) also recently completed a Display Stream Compression (DSC) standard for next-generation mobile or TV/Computer display interfaces.

The development of these standards introduced many new ideas, which are expected to inspire more future innovations and benefit the varied usage of screen content coding.

In contrast with other image or video compression standards, the proposed Display Stream Compression standard targets a relatively low compression ratio and emphasizes visually-lossless performance, high data throughput, low latency, low complexity, and includes special considerations geared for future display architectures.

VESA's DSC standard version 1.0 enables up to 66% data rate reduction, extending battery life in mobile systems and laptops, while simplifying the electrical interface requirements for future 4K and 8K displays. The standard enables a single codec for system chips that have multiple interfaces.

As display resolutions continue to increase, the data rate across the video electrical interface has also increased. Higher display refresh rates and color depths push rates up even further. For example, a 4K display at 60 frames per second with a 30 bit color depth requires a data rate of about 17.3 gigabits per second, which is the current limit of the Display Port specification. Higher interface data rates demand more power, can increase the interface wire count, and require more shielding to prevent interference with the device's wireless services. These attributes increase system hardware complexity and weight and are undesirable for today's sleek product designs.

http://www.prweb.com/releases/real-time-video/compression-VESA-DSC/prweb11732917.htm

Implement the DSC standard using various test sequences. (www.vesa.org) For non VESA members the DSC standard is available for $350.00.

P.5.130 Yan et al. [E221] have achieved significant speed up by implementing efficient parallel framework for HEVC motion estimation on multi core processors compared with serial execution. In the conclusions they state that they like to find efficient parallel methods for other processing stages in the encoder and to find an efficient parallel framework for HEVC encoder (Figure 5.5). Explore this.

P.5.131 See [E223] This can lead to several projects. Several papers related to frame recompression are listed in the references. Pixel truncation scheme for motion estimation [17], combining pixel truncation and compression to reconstruct lossless pixels for MC [9, 18], new frame recompression algorithm integrated with H.264/AVC video compression [7], See also [9] and [11]. Also the mixed lossy and lossless reference frame recompression is proposed in [E223]. These techniques can be investigated for improved compression in HEVC and as well ME/MC functions. Both [E223] and the papers listed as references need to be reviewed thoroughly.

P.5.132 See P.5.131. The frame recompression techniques proposed and applied to HEVC in [E223] (besides H.264/AVC) can also be explored in other video coding standards such as DIRAC, AVS China, VP9 and VC-1. Each of these can be a project/thesis. Investigate in detail with regards to additional gains that can be achieved by integrating frame recompression with these standards.

P.5.133 The papers presented in IEEE ICCE 2015 (Las Vegas, NV, Jan. 2015) are listed in [E224–E238]. In IEEE ICCE, in general, each paper (extended abstracts) is limited to 2 pages. The authors can be contacted by emails and full papers, if any, can be requested. These papers can lead to additional projects.

P.5.134 Zhao, Onoye and Song [E240] have developed detailed algorithms for HEVC resulting in 54.0–68.4% reduction in encoding time with negligible performance degradation. They also show that fast intra mode decision algorithm can be easily implemented on a hardware platform with fewer resources consumed. Simulation results are based on class A through class E test sequences using HM 11. Implement these algorithms using HM 16 and extend to 8K $\times$ 4K sequences.

P.5.135 See P.5.134. The authors [E240] suggest some aspects of future work in conclusions. Explore these in detail with the goal of reducing the HEVC encoder complexity even further.

P.5.136 P.5.136 you, Chang and Chang [E239] have proposed an efficient ME design with a joint algorithm and architecture optimization for HEVC encoder. In Table XV, this proposal is compared with other ME designs and demonstrate that their design reduces the gate count and on-chip memory size significantly. Verify their ME design using HM 16.0 and extend to 8K $\times$ 4K sequences.

P.5.137 The theory behind development of DCT based fractional pel interpolation filters and their selection in HEVC is described clearly in [E109]. The performance of these interpolation filters in HEVC is compared with corresponding filters adopted in H.264/AVC. Also some of the fractional pel interpolation filters in H.264/AVC are replaced with those in HEVC to evaluate their effects. Go through this paper and the related references listed at the end and confirm the results shown in Tables 3–6 and Figure 6. Use HM 16.0. Replace the interpolation filters adopted in H.264/AVC completely by those specified in HEVC and evaluate its performance.

P.5.138 Umezaki and Goto [E241] have developed two methods for region of interest based streaming in HEVC and have evaluated in terms of decoding cost, bandwidth efficiency and video quality. They suggest possible future research directions to further reduce the decoding cost. Explore these directions further to achieve the desired results.

P.5.139 Correa et al. [E242] have achieved an average complexity reduction of 65% by combining three early termination schemes in CTU, PU and RQT structures with a negligible BD-rate increase of 1.36% in the HEVC encoder. Their proposed method uses data mining as a tool to build a set of decision trees that allow terminating the decision processes thus relieving the encoder of testing all encoding structure partitioning possibilities. These results are confirmed by simulations based on various test sequences (different resolutions) and are compared with earlier works. Implement this work and confirm their results. Extend this to 4K $\times$ 2K and 8K $\times$ 4K test sequences. Can the HEVC encoder complexity reduction be further Improved?

P.5.140 Ngyuen and Marpe [E245] have conducted a detailed performance comparison of HEVC main still picture (MSP) profile with other still image and video compression schemes (intra coding only) such as JPEG, JPEG 2000, JPEG XR, H.264/MPEG-4 AVC, VP8, VP9 and WebP. They used PSNR and BD-bit rate as the comparison metrics (see Figures 2–4 in [E245]). Extend this comparison using test sequences with spatial resolution higher than 1280 $\times$ 1600. Another image compression standard is JPEG-LS. Consider this also regarding comparison with HEVC MSP profile.

P.5.141 See ([E245] and P.5.140). Consider mean opinion score (MOS) as the metric for performance comparison of all these standards. Access reference 35 in [E245] regarding HEVC subjective evaluation. Note that this comparison requires extensive man power, detailed testing lab and other resources.

P.5.142 Min and Cheung [E247] have developed a fast CU decision algorithm that reduces the HEVC intra encoder complexity by nearly

52% with negligible R-D performance loss. In the conclusion section they state that for the blur sequences that the BD rate [E81, E82, E96, E198] is worse than the sequences with sharp edges. Explore how this can be minimized for sequences with blur background.

P.5.143 Kim and Park [E248] have proposed a CU partitioning method that reduces the HEVC encoder complexity by 53.6% on the average compared with the HM 15.0 with negligible coding efficiency loss. They state that further work will focus on the optimal feature selection for CU partitioning. Explore and implement this selection.

P.5.144 Chi et al. [E246] developed SIMD optimization for the entire HEVC decoder resulting in 5x speed up. This has been substantiated by implementing the optimized HEVC decoder on 14 mobile and PC platforms covering most major architectures released in recent years. In the conclusions, they state "General purpose architectures have lately increased their floating point performance at a much faster rate than integer performance and in the latest architectures have even a higher floating point throughput. As the use of floating point numbers might also improve compression performance, it is an interesting and promising direction for future work". Implement SIMD acceleration for HEVC decoding using floating point arithmetic and evaluate any improvement in compression performance. This project is complex and may require a group of researchers.

P.5.145 Hautala et al. [E253] have developed low-power multicore coprocessor architecture for HEVC/H.265 in-loop filtering that can decode 1920 × 1080 video sequences (luma only) at 30 fps. Extend this architecture for in-loop filtering of chroma components also. They state that the in-loop filters (both deblocking and SAO) typically consume about 20% of the total decoding time.

P.5.146 See [E254]. Go through this paper in detail. Extend this scheme to 4K and 8K video sequences using the latest HM. Evaluate the bit-rate reductions, subjective quality and computational complexity and compare with the conclusions in [E254].

P.5.146 See P.5.145. The authors state that by using four instances of the proposed architecture, in-loop filtering can be extended to 4K sequences. Explore and implement this.

P.5.147 Zhang et al. [E259] have proposed a Machine Learning-Based Coding Unit depth decision that reduces the HEVC encoder complexity on average 51.45% compared with very little loss in BD-bit rate and BD-PSNR. Go through this paper in detail and extend this technique to 4K and 8K test sequences. Compile a comparative table similar to Table III in this paper.

P.5.148 Chen et al. [E257] have developed a New Block Based Coding method for HEVC Intra Coding, which results in a 2% BD-rate reduction on average compared with the HM 12.0 Main Profile Intra Coding. However, the encoding time has increased by 130%. In terms of future work, the authors plan on designing new interpolators to improve the R-D performance and also extend this technique to inter coding. Explore this in detail. Consider various avenues related to reducing the encoder complexity (130% increase in the complexity essentially nullifies any BD-rate reductions). Consider the latest HM.

P.5.149 Zou et al. [E255] have developed a Hash Based Intra String Copy method for HEVC based Screen Content Coding that achieves 12.3% and 11% BD-rate savings for 1080P RGB and YUV respectively in full frame Intra Block Copy condition. However, this reduction in BD rates comes at the cost of 50% increase in encode complexity (see Table 1). Can the encoder complexity be reduced by disabling the proposed mode for certain CU sizes. Explore this in detail.

P.5.150 Hu et al. [E258] have developed a Hardware-Oriented Rate-Distortion optimization Algorithm for HEVC Intra-Frame Encoder that achieves 72.22% time reduction of rate-distortion optimization (RDO) compared with original HEVC Test Model while the BD-rate is only 1.76%. Also see all the references listed at the end in [E258]. They suggest that the next step is to implement in Hardware. Design and develop an encoder architecture layout, to implement the same.

P.5.151 In [E261] several methods such as mode mapping, machine learning, complexity scalable and background modeling for H.264/AVC to HEVC/H.265 transcoder are explained. Implement all these techniques for H.264/AVC to HEVC transcoder and compare their performances using standard metrics. See references at the end.

P.5.152 E. Peixoto [E263] et al. have developed a Fast H.264/AVC to HEVC Transcoding based on Machine Learning, which is 3.4 times faster, on average, than the trivial transcoder, and 1.65 times faster than a previous transcoding solution. Explore HEVC to H.264/AVC transcoder based on Machine Learning for various Profiles/Levels.

P.5.153 Please access the paper, D. Mukherjee, "An overview of new video coding tools under consideration for VP10: the successor to VP9,"

[9599 – 50], SPIE. Optics + photonics, San Diego, California, USA, 9–13, Aug. 2015. Implement VP10 encoder using new video coding tools under consideration and compare with HEVC (H.265) based on standard metrics.

P.5.154 Please access the paper, D. Mukherjee, "An overview of new video coding tools under consideration for VP10: the successor to VP9," [9599 – 50], SPIE. Optics + photonics, San Diego, California, USA, 9–13, Aug. 2015. Develop the VP10 decoder block diagram, compare its implementation complexity with HEVC decoder.

P.5.155 Please see the Paper, R. G. Wang et al., "Overview of MPEG Internet Video Coding", [9599 – 53], SPIE. Optics + photonics, San Diego, California, USA, 9–13, Aug. 2015 and access all the references listed at the end. In response to the call for proposals for internet video coding by MPEG, three different codecs Web Video Coding (WVC), Video Coding for browsers (VCB) and Internet Video Coding (IVC). WVC and VCB were proposed by different group of companies and IVC was proposed by several Universities and its coding tools were developed from zero. Section 2 describes coding tools (key technologies) used in the current test model of IVC (ITM 12.0). Specific Internet applications and the codec requirements are listed. Besides an overview of IVC, this paper presents a performance comparison with the WVC and VCB codecs using different test sequences at various bit rates (Table 9). Constraints are listed in Section 3.1—Test cases and Constraints. Implement the IVC, WVC and VCB codecs and compare with AVC HP (H.264). Extend this comparison based on 4k × 2k video test sequences (Table 8). Consider also implantation complexity as another comparison metric. Implementation complexity requirements are also described in this paper (Section 1). Please see [38–41] in references for detailed encoding settings.

Note that JVT-VC documents can be accessed as follows:

E1. JCT-VC DOCUMENTS can be found in JCT-VC document management system
http://phenix.int-evry.fr/jct (see [E185])

E2. All JCT-VC documents can be accessed. [online]. Available: http://phenix.intevry.fr/jct/doc_end_user/current_meeting.php?id_meeting=154&type_order=&sql_type=document_number

P.5.156 Please see the Paper, R. G. Wang et al., "Overview of MPEG Internet Video Coding", [9599 – 53], SPIE. Optics + photonics, San Diego, California, USA, 9–13, Aug. 2015. This Paper says "VCB was proposed by Google and it is in fact VP8". Replace VP8 by the tools proposed in VP10 and implement the VCB in detail and evaluate thoroughly its performance with IVC, WVC and AVC HP. Please refer the paper, D. Mukherjee, "An overview of new video coding tools under consideration for VP10: the successor to VP9," [9599 – 50], SPIE. Optics + photonics, San Diego, California, USA, 9–13, Aug. 2015.

P.5.157 Please review [E266] in detail. This paper has proposed a Novel JND-based HEVC – Complaint Perceptual Video Coding (PVC) scheme that yielded a remarkable bitrate reduction of 49.10% maximum and 16.10% average with negligible subjective Quality loss. Develop and implement A VP10 – Compliant Perceptual Video Coding Scheme based on JND models for Variable Block – sized Transform Kernels and consider bit-rate reductions, implementation complexity, subjective quality etc., as performance metrics. Subjective Quality evaluation requires extensive test set up (Monitor, lighting, and display) and a number of trained viewers to get Mean Opinion Score (MOS). (Also refer: D. Mukherjee, "An overview of new video coding tools under consideration for VP10: the successor to VP9," [9599 – 50], SPIE. Optics + photonics, San Diego, California, USA, 9–13, Aug. 2015 and Ref 11 at the end of [E266]).

P.5.158 Please see P.5.157. Repeat this project for AVS – China. (See References in AVS China Section).

P.5.159 Please see P.5.157. Repeat this project for DIRAC. (See References in DIRAC Section).

P.5.160 Please see P.5.157. Repeat this project for HEVC - IVC. (Also Refer: R. G. Wang et al., "Overview of MPEG Internet Video Coding", [9599 – 53], SPIE. Optics + photonics, San Diego, California, USA, 9–13, Aug. 2015).

P.5.161 Review [E264]. This paper presents an overview of scalable extensions of HEVC where in the scalabilities include Spatial, SNR, bit depth and color gamut as well as combinations of any of these. In the conclusions, it is stated that ATSC is considering SHVC for broadcasting and video streaming. Go to www.atsc.org and extract related documents. Implement SHVC for different scalabilities compatible with ATSC requirements. Go to http://atsc.org/newsletter/2015, to get newsletters on different video coding standards.

P.5.162 Review [E264]. Implement HEVC (all profiles/levels) compatible with ATSC requirements. Go to www.atsc.org. Also refer to P.5.161. Go to http://atsc.org/newsletter/2015, to get newsletters on different video coding standards.

P.5.163 Chen et al. [E267] have proposed a novel framework for software based H.264/AVC to HEVC transcoding, integrated with parallel processing tools that are useful for achieving higher levels of parallelism on multicore processors and distributed systems. This proposed transcoder can achieve up to 60x speed up on a Quad core 8-thread server over decoding—re-encoding based on FFMPEG and the HM software with a BD-rate loss of 15–20% and cam also achieve a speed of 720P at 30 Hz by implementing a group of picture level task distribution on a distributed system with nine processing units. Implement the same for higher resolution (8k $\times$ 4k) sequences.

P.5.164 Please see P.5.163. Develop a framework for software based HEVC (All Profiles/Levels) to H.264/AVC transcoding.

P.5.165 Please see P.5.163. Develop an algorithm for H.264 to HEVC transcoding that can reduce compression performance loss while improving/maintaining transcoding speed for 4k $\times$ 2k video sequences. Also extend to 8k $\times$ 4k video sequences.

P.5.166 Lee et al. [E268] have developed an early skip mode decision for HEVC encoder that reduces the encoder complexity by 30.1% for random access and around by 26.4% for low delay with no coding loss. Implement this for 4k and 8K video using the latest HM software.

P.5.167 Lim et al. [E269] have developed a fast PU skip and split termination algorithm that achieves a 44.05% time savings on average when α = 0.6 and 53.52% time savings on average when α = 0.5 while maintaining almost the same RD performance compared with HM 14.0. In the conclusion it is stated that the there are some complexity and coding efficiency tradeoff differences between class F and other class sequences in the simulation results. Explore this as future research.

P.5.168 Won and Jeon [E270] have proposed a complexity – efficient rate estimation for mode decision of the HEVC Encoder that shows an average saving in rate calculation time for all intra, low delay and random access conditions of 55.29%, 41.05% and 42.96% respectively. Implement this for 4k and 8k video sequences using the latest HM software.

P.5.169 Chen et al. [E272] have developed a novel wavefront based high parallel (WHP) solution for HEVC encoding integrating data level and task level methods that bring up to 88.17 times speedup on 1080P sequences, 65.55 times speed up on 720P sequences and 57.65 times speedup on WVGA sequences compared with serial implementation. They also state that the proposed solution is also applied in several leading video companies in China, providing HEVC video service for more than 1.3 million users every day. Implement this for 4k and 8k video using the latest HM software. Develop tables similar to Tables III through XI and graphs similar to Figures 14 and 15. If possible use the hardware platform described in Table IV.

P.5.170 Zhang, Li and Li [E273] have proposed an efficient fast mode decision method for Inter Prediction in HEVC which can save about 77% encoding time with only about 4.1% bit rate increase compared with HM16.4 anchor and 48% encoding time with only about 2.9% bit rate increase compared with fast mode decision method adopted in HM16.4. Flow chart of the proposed algorithm is shown in Figure 5. Previous Fast Mode Decision Algorithms designed for HEVC Inter prediction are also described. Review the paper. Implement the same and confirm the results. If possible use hardware platform described in Table IV.

P.5.171 Please see P.5.170. Implement the code based on the given flow chart using C++ compiler and use "clock" function to measure the run time.

P.5.172 Please see P.5.170. Implement the same project for 4k and 8k video sequences and tabulate the results.

P.5.173 Hu and Yang [E274] have developed a Fast Mode Selection for HEVC intra frame coding with entropy coding refinement. Using various test sequences, they have achieved 50% reduction in encoding time compared with HM 15.0 with negligible BD-Rate Change. Review this paper thoroughly and confirm their results. Test sequences for the simulation are class A through class F (see Table IV). Extend this simulation using 4k and 8k sequences and cite the results and conclusions.

P.5.174 Jou, Chang and Chang [E275] have developed an efficient ME design with a joint algorithm and architecture optimization that reduces the integer Motion Estimation and fractional Motion Estimation complexity significantly. Architectural design also supports real-time encoding of 4k $\times$ 2k video at 60 fps at 270] MHZ. Review this paper in detail and simulate the algorithm/architecture Motion Estimation design. Please confirm the results shown in the last column of Table XV. In the conclusion it says further optimization can be achieved by tweaking the proposed algorithms and corresponding architecture. Explore this in detail.

P.5.175 Go through the overview paper on emerging HEVC SCC extension described in [E322] and the related references listed at the end. Using the test sequences as described in Tables III and IV, verify the results shown in Table V using JM 18.6 (H.264/AVC), HM 16.4 (HEVC) and SCM 3.0/SCM 4.0 (HEVC–SCC).

P.5.176 Repeat P.175 using Table VI from the overview paper on emerging HEVC SCC extension described in [E322].

P.5.177 Repeat P.175 using Table VII from the overview paper on emerging HEVC SCC extension described in [E322].

P.5.178 Repeat P.175 using Table VIII from the overview paper on emerging HEVC SCC extension described in [E322].

P.5.179 Repeat P.175 using Table XI from the overview paper on emerging HEVC SCC extension described in [E322].

P.5.180 Repeat P.175 using Table XII from the overview paper on emerging HEVC SCC extension described in [E322].

P.5.181 Repeat P.175 using Table XV from the overview paper on emerging HEVC SCC extension described in [E322].

P.5.182 Repeat P. 175 using Table XVI from the overview paper on emerging HEVC SCC extension described in [E322].

Note that the projects P.5.175 through P.5.182 require thorough understanding of the H.264/AVC, HEVC version 1 and HEVC–SCC besides familiarity of the corresponding software JM 18.6, HM 16.4 and SCM 3.0/SCM 4.0 (respectively). Note that implementation complexity is not considered in any of these simulations. The additional tools/modules added to the HEVC version for SCC as described in [E322] probably result in increased complexity. This is a relevant comparison metric. Develop Tables showing the comparison of this complexity.

P.5.183 Using standard test sequences, confirm the results shown in Table II [E321] for all intra (AI), RA (Random Access) and LD (Low Delay) configurations.

P.5.184 Repeat P.5.183 for lossless performance. See Table III [E321].

P.5.185 Repeat P.5.184. Extend the performance comparison of version II with H.264/AVC (see Tables II and III) [E321] based on implementation complexity as a metric.

P.5.186 Using the test sequences in Table 5 [E325], confirm the results shown in Table 6 [E325] in terms of coding performance of SHVC and HEVC simulcast.

P.5.187 See P.5.186. Confirm the results shown in Table 7 [E325] in terms of coding performance of SHVC EL and HEVC single-layer coding equivalent to EL.

P.5.188 See P.5.186. Confirm the results shown in Table 8 [E325] in terms of coding performance of SHVC and HEVC single-layer coding equivalent to EL.

P.5.189 See P.5.186. Confirm the results shown in Table 9 [E325] in terms of coding performance of SHVC and SVC.

P.5.190 In [E325] implementation complexity has not been included as a performance metric. Evaluate this for all cases listed in Tables 6 thru 8 [E325].

P.5.191 [E348] M.S. Thesis "Reducing encoder complexity of intra-mode decision using CU early termination algorithm' by Nishit Shah. By developing CU early termination and TU mode decision algorithm, he has reduced the computational complexity for HEVC intra prediction mode with negligible loss in PSNR and slight increase in bit rate." Conclusions and future work are reproduced here:

Conclusions

In this thesis a CU splitting algorithm and the TU mode decision algorithm are proposed to reduce the computational complexity of the HEVC encoder, which includes two strategies, i.e. CU splitting algorithm and the TU mode decision. The results of comparative experiments demonstrate that the proposed algorithm can effectively reduce the computational complexity (encoding time) by 12–24% on average as compared to the HM 16.0 encoder [38], while only incurring a slight drop in the PSNR and a negligible increase in the bitrate and encoding bitstream size for different values of the quantization parameter based on various standard test sequences [29]. The results of simulation also demonstrate negligible decrease in BD-PSNR [30], i.e., 0.29–0.51 dB as compared to the original HM16.0 software and negligible increase in the BD-bitrate [31].

Future Work

There are many other ways to explore in the CU-splitting algorithm and the TU mode decision in the intra prediction area. Many of these methods can be combined with this method, or if needed, one method may be replaced by a new method and encoding time gains can be explored.

Similar algorithms can be developed for fast inter-prediction in which the RD cost of the different modes in inter-prediction are explored, and depending upon the adaptive threshold, mode decision can be terminated resulting in less encoding time and reduced complexity combining with the above proposed algorithm.

Another fact of encoding is CU size decisions which are the leaf nodes of the encoding process in the quad tree. Bayesian decision rule can be applied to calculate the CU size and then this information can be combined with the proposed method to achieve further encoding time gains.

Complexity reduction can also be achieved through hardware implementation of a specific algorithm which requires much computation. The FPGA implementation can be useful to evaluate the performance of the system on hardware in terms of power consumption and encoding time. Explore this Future Work.

P.5.192 Hingole [E349] has implemented the HEVC bit stream to H.264/MPEG4 AVC bit stream transcoder (see figure below)

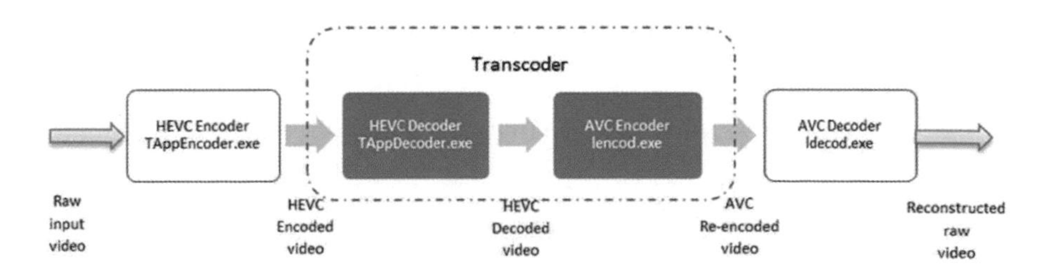

The transcoder can possibly be significantly reduced in complexity by adaptively reusing the adaptive intra directional predictions (note HEVC has 35 intra directional predictions versus 9 for H.264/MPEG4 AVC) and MV reuse from HEVC in H.264 (again HEVC multiple block sizes, both symmetric and asymmetric, for ME and up to 1/8 fractional MV resolution versus simpler block sizes in H.264). Explore this in detail and see how the various modes in HEVC can be tailored to those in H.264/AVC such that the overall transcoder complexity can be significantly reduced. This requires thorough understanding of both HEVC and H.264 codecs besides review of the various transcoding techniques already developed.

P.5.193 Dayananda [E350] in her M.S. thesis entitled *"Investigation of Scalable HEVC & its bitrate allocation for UHD deployment in the context of HTTP streaming", has suggested*
Further, work on exploring optimal bitrate algorithms for allocation of bits into layer of SHVC based on Game theory and other approaches can be done, considering various scalability options (such as spatial, quality and combined scalabilities). Explore this.

P.5.194 See P.5.192 Additional experiments to study the effect of scalability overhead for its modeling can be done using several test sequences. Implement this.

P.5.195 See P.5.192 Also, evaluation of SHVC for its computational complexity can be done and parallel processing techniques for encoding base and enhancement layers in SHVC can be explored. Investigate this. Note that this is a major task.

P.5.196 See [E351] M.S. thesis by V. Vijayaraghavan, "Reducing the encoding time of motion estimation in HEVC using parallel programming". Conclusions and future work are reproduced here;

Conclusions

Through thorough analysis with the most powerful tool, Intel® vTune™ amplifier, hotspots were identified in the HM16.7 encoder. These hotspots are the most time consuming functions/loops in the encoder. The functions are optimized using optimal C++ coding techniques and the loops that do not pose dependencies are parallelized using the OpenMP directives available by default in Windows Visual Studio.

Not every loop is parallelizable. Thorough efforts are needed to understand the functionality of the loop to identify dependencies and the capability of the loop to be made parallel. Overall observation is that the HM code is already vectorized in many regions and hence parallel programming on top of vectorization may lead to degradation in performance in many cases. Thus the results of this thesis can be summarized as below:

- Overall ~24.7–42.3% savings in encoding time.
- Overall ~3.5–7% gain in PSNR.
- Overall ~1.6–4% increase in bitrate.

Though this research has been carried out on a specific configuration (4 core architecture), it can be used on any hardware universally. This implementation works on servers and Personal Computers. Parallelization in this thesis has been done at the frame level.

Future Work:

OpenMP framework is a very simple yet easy to adapt framework that aids in thread level parallelism. Powerful parallel programming APIs are available which can be used in offloading the serial code to the GPU. Careful efforts need to be invested in investigating the right choice of software and functions in the software chosen to be optimized. If optimized appropriately, huge savings in encoding time can be achieved.

Intel® vTune™ amplifier is a very powerful tool which makes it possible for analysis of different types to be carried at the code level as well as at the hardware level. The analysis that has been made use of in this thesis is Basic Hotspot analysis. There are other options available in the tool, one of which helps us to identify the regions of the code which cause the maximum number of locks and waits and also the number of cache misses that occur. Microprocessor and assembly level optimization of the code base can be achieved by diving deep into this powerful tool.

See [E351], parallel programming of CPU threads is achieved using OpenMP parallel threads. More efficient times can be obtained (reduced encoding time) using GPU (Graphics Processing Unit) programming with OpenCL framework. Explore this.

Intel vTune Amplifier basic hotspot analysis has been used to identify the top hotspots (functions and loops) in the code using Basic Hotspot Analysis which is used for code optimization. There are several different analysis options available in IntelvTune Amplifier, which can be used to optimize the code at the assembly level. Assembly level optimizations can further increase the efficiency of the codec and reduce the encoding time.

Investigate the suggested optimizations based on these criteria.

P.5.197 Kodpadi [E352] in her M.S. thesis (EE Dept., UTA) entitled "Evaluation of coding tools for Screen content in High-Efficiency Video Coding", has suggested:
The future work can be on reducing the encoding time by parallelizing the different methods or by developing fast algorithms on the encoder side. Explore this.

P.5.198 See P.5.197 Continue to evaluate the coding performance of the newly adopted tools and their interaction with the existing HEVC tools in the Main profile and range extensions. See OP5 and OP7.

P.5.199 See P.5.197 Study latency and parallelism implications of SCC coding techniques, considering multicore and single-core architectures.

P.5.200 see P.5.197 Analyze complexity characteristics of SCC coding methods with regards to throughput, amount of memory, memory bandwidth, parsing dependencies, parallelism, pixel processing, Chroma position interpolation, and other aspects of complexity as appropriate.

P.5.201 Mundgemane [E353] in her M.S. thesis entitled "Multi-stage prediction scheme for Screen Content based on HEVC", has suggested:
The future work can be on reducing the encoding time by parallelizing the independent methods on the encoder side. Implement this.

P.5.202 See P.5.201 Implement fast algorithms for screen content coding tools that can help in decreasing the encoding time.
P.5.203 See P.5.201 The coding efficiency of HEVC on surveillance camera videos can be explored. Investigate this.

P.5.203 See [H69]. This paper describes several rate control algorithms adopted in MPEG-2, MPEG-4, H.263 and H.264/AVC. The authors propose a rate control mechanism that significantly enhances the RD performance compared to the H.264/AVC reference software (JM 18.4), Several interesting papers related to RDO of rate control are cited at the end. Investigate if the rate control mechanism proposed by the authors can be adopted in HEVC and compare its performance with the latest HM software. Consider various test sequences at different bit rates. Develop figures and tables similar to Figures 8–10 and Tables I to III shown in [H69] for the HEVC. This project is extensive and can be applied to M.S. Thesis.

P.5.204 See P.5.203. In [H69], computational complexity has not been considered as a performance metric. Investigate this. It is speculated that the proposed rate control mechanism invariably incurs additional complexity compared to the HM software.

P.5.205 Hamidouche, Raulet and Deforges [E329] have developed a software parallel decoder architecture for the HEVC and its various multilayer extensions. Advantages of this optimized software are clearly explained in the conclusions. Go through this paper in detail and develop similar software for the HEVC decoder for scalable and multiview extensions.

P.5.206 Kim et al. [E355] have implemented the cross component prediction (CCP) in HEVC (CCP scheme is adopted as a standard in range extensions) and have shown significant coding performance improvements for both natural and screen content video. Note that the chroma residual signal is predicted from the luma residual signal inside the coding loop. See Figure 2 that shows the block diagrams of encoder and decoder with CCP. This scheme is implemented in both RGB and YCbCr color spaces.
In the conclusions, the authors state that more study can be made on how to facilitate software and hardware implementation. Investigate this.

P.5.207 See P.5.206 The authors also state "We leave it as a further study how to apply CCP to various applications including HDR image coding." Investigate this.

P.5.208 Fong, Han and Cham [E356] have developed recursive integer cosine transform (RICT) and have demonstrated from order-4 to order-32. The proposed RICT is implemented into reference software HM 13.0. By using different test sequences, they show that RICT has similar coding efficiency (see Tables VI thru XI in this paper) as the core transform in HEVC. See references 18 thru 22 listed in [E356]. Using the recursive structure develop order-64 and order-128 RICTs and draw the flow graphs similar to Figure 1 and the corresponding matrices similar to Equation 16. The higher order transforms are proposed in beyond HEVC [BH1, BH2].

P.5.209 Hsu and Shen [E357] have designed a deblocking filter for HEVC that can achieve 60 fps for the video with 4K $\times$ @2K resolution assuming an operating frequency of 100 MHz. Go through the VLSI architecture and hardware implementation of this filter and extend the design so that the deblocking filter can operate on the video with 8K $\times$ 4K resolution. This requires major hardware design tools.

P.5.210 Francois et al. [E339] present the roles of high dynamic range (HDR) and wide color gamut (WCG) in HEVC in terms of both present status and future enhancements. They describe in detail the various groups involved in HDR/WCG standardization including the work in MPEG and JCT-VC. They conclude that the groups' efforts involve synchronization to ensure a successful and interoperable market deployment of HDR/WCG video. The proposals on HDR/WCG video coding submitted to JCT-VC can lead to several projects. The references related to these proposals are listed at the end in [E339]. In the conclusions the authors state "Finally following the conclusions of the CfE for HDR and WCG video coding, MPEG has launched in June 2015 a fast-track standardization process to enhance the performance of the HEVC main 10 profile for HDR and WCG video, that would lead to an HEVC extension for HDR around mid-2016." (CfE: Call for evidence). Figure 6 shows the application of CRI (color remapping info.) for the HDR and WCG to SDR conversion. Implement this and show the two displays (SDR and HDR/WCG).

P.5.211 See P.5.210. Two methods for conversion from 8 bit BL resolution BT.709 to 10 bit EL resolution BT.2020 are shown in Figure 10. The authors state that the second method (Figure 10(b)) can keep the conversion precision and has better coding performance compared with the first method (Figure 10(a)). Implement both methods and confirm this comparison.

P.5.212 See P.5.210 Another candidate for HDR/WCG support is shown in Figure 11. Using a 16 bit video source, split into two 8 bit videos followed by two legacy HEVC coders and combining into 16 bit reconstructed video. Implement this scheme and evaluate this in terms of standard metrics.

P.5.213 See P.5.210 An alternate solution for HDR video coding is illustrated in Figures 12 and 13. Simulate this scheme and compare with the techniques described in P.5.210–P.5.212.

P.5.214 See P.5.210 HDR/WCG video coding in HEVC invariably involves additional computational complexity. Evaluate the increase in complexity over the HEVC coding without these extensions. See [E84].

P.5.215 Tan et al. [SE4] have presented the subjective and objective results of a verification test in which HEVC is compared with H.264/AVC. Test sequences and their parameters are described in Tables II and III and are limited to random access (RA) and low delay (LD).

Extend these tests to all intra (AI) configuration and develop figures and tables as shown in Section IV Results. Based on these results confirm the bit rate savings of HEVC over H.264/AVC.

P.5.216 Lee et al. [E358] have developed a fast RDO quantization for HEVC encoder that reduces the quantization complexity for AI, RA and LD with negligible coding loss. See Tables X thru XIII. They propose to include a fast context update with no significant coding loss. Explore this. Develop tables similar to these specifically showing further reduction in complexity.

P.5.217 Georgios, Lentaris and Reisis [E360] have developed a novel compression scheme by sandwiching the traditional video coding standards such as H.264/VAC and HEVC between down sampling at the encoder side and upsampling at the decoder side. They have shown significant encoder/decoder complexity reduction while maintaining the same PSNR specifically at low bitrates. The authors have proposed SR (super resolution) Interpolation algorithm called L-SEABI (low-complexity back-projected interpolation) and compared their algorithms with state-of-the-art algorithms such as bicubic interpolation based on various test sequences in terms of PSNR, SSIM and BRISQUE (blind/reference less image spatial quality evaluator) (See Table I). Implement all the interpolation algorithms and verify Table I.

P.5.218 See P.5.217. Please go through the conclusions in detail. The authors state that their SR compression scheme out performs the H.264/HEVC codecs for bitrates up to 10 and 3.8 Mbps respectively, consider BD bitrate and BD-PSNR [E81, E82, E96, E198] as comparison metrics. Include these in all the simulations and develop tables similar to Figure 7.

P.5.219 See P.5.218. Implement this novel SR compression scheme in VP9 and draw the conclusions.

P.5.220 See P.5.219. Replace VP9 by AVS China and implement the SR compression scheme.

P.5.221 Repeat P.5.220 for DIRAC codec developed by BBC.

P.5.222 Repeat P.5.220 for VC1 codec developed by SMPTE.

P.5.223 Kuo, Shih and Yang [E361] have improved the rate control mechanism by adaptively adjusting the Lagrange parameter and have shown that their scheme significantly enhances the RD performance compared to the H.264/AVC reference software. Go through this paper and all related references in detail.

Can a similar scheme be applied to HEVC reference software. If so, justify the improvement in RD performance by applying the adaptive scheme in various test sequences (See corresponding Figures and Tables in [E361]).

P.5.224 Jung and Park [E332] by using an adaptive ordering of modes have proposed a fast mode decision that significantly reduces the encoding time for RA and LD configurations. Can you extend this scheme to AI case. If so, list the results similar to Tables XI, XII and XIII based on the test sequences outlined in Table IX followed by conclusions. Use the latest HM software.

P.5.225 Zhang, Li and Li [E362] have developed a fast mode decision for inter prediction in HEVC resulting in significant savings in encoding time compared with HM16.4 anchor. See Tables VI thru XII and Figure 9. The test sequences are limited to 2560 × 1600 resolution. Extend the proposed technique to 4K and 8K video sequences and develop similar Tables and Figure. Based on the results, comment on the corresponding savings in encoder complexity.

P.5.226 Fan, Chang and Hsu [E363] have developed an efficient hardware design for implementing multiple forward and inverse transforms for various video coding standards. Can this design be extended to HEVC also? Investigate this in detail? Note that in HEVC 16 × 16 and 32 × 32 INTDCTs are also valid besides 4 × 4 and 8 × 8 INTDCTs.

P.5.227 Park et al. [E364] have proposed a 2-D 16 × 16 and 32 × 32 inverse INTDCT architecture that can process 4K@30fps video. Extend this architecture for 2-D inverse 64 × 64 INTDCT that can process 8K@30 fps video. Note that 64 × 64 INTDCT is proposed as a future development in HEVC (beyond HEVC). (See the papers in BEYOND HEVC).

P.5.228 Au-Yeung, Zhu and Zeng [H70] have developed partial video encryption using multiple 8 × 8 transforms in H.264 and MPEG-4. Can this encryption approach be extended to HEVC. Note that multiple size transforms are used in H.265. Please access the paper by these authors titled “Design of new unitary transforms for perceptual video encryption” [H71].

P.5.229 See [H72]. Can the technique of embedding sign-flips into integer-based transforms be applied to encryption of H.264 video be extended to H.265. Explore this in detail. See also [H73].

P.5.230 See [E365] The authors have proposed a collaborative inter-loop video encoding framework for CPU+GPU platforms. They state "the proposed framework outperforms single-device executions for several times, while delivering the performance improvements of up to 61.2% over the state of the art solution." This is based on the H.264/AVC inter-loop encoding. They also state extending this framework to other video coding standards such as HEVC/H.265 as future work. Explore this future work in detail.

P.5.231 See [Tr11] The authors have presented an approach for accelerating the intra CU partitioning decision of an H.264/AVC to HEVC transcoder based on Naïve-Bayes classifier that reduces the computational complexity by 57% with a slight BD-rate increase. This is demonstrated by using classes A thru E test sequences (see Table 1). Go through this paper in detail and confirm their results. Extend these simulations to 4K and 8K test sequences and verify that similar complexity reduction can be achieved.

P.5.232 See [E191] and P.5.109. Extend the comparative assessment of HEVC/H.264/VP9 encoders for random access and all intra applications using 1280 × 720 @60 fps test sequences. Extend this to UHDTV test sequences. Consider also implementation complexity as another metric.

P.5.233 Jridi and Meher [E367] have derived an approximate kernel for the DCT of length 4 from that defined in HEVC and used this for computation of DCT and its inverse for power-of-2 lengths. They discuss the advantages of this DCT in terms of complexity, energy efficiency, compression performance etc. Confirm the results as shown in Figures 8 and 9 and develop the approximate DCT for length 64.

P.5.234 See [E368] Section VII Conclusion is reproduced here.

"In this paper, fast intra mode decision and CU size decision are proposed to reduce the complexity of HEVC intra coding while maintaining the RD performance. For fast intra mode decision, a gradient-based method by using average gradients in the horizontal (AGH) and vertical directions (AGV) is proposed to reduce the candidate modes for RMD and RDO. For fast CU size decision, the homogenous CUs are early terminated first by employing AGH and AGV. Then two linear SVMs which employ the depth difference, HAD cost ratio (and RD cost ratio) as features are proposed to perform early CU split decision and early CU termination decision for the rest of CUs. Experimental results show that the proposed method achieves a significant encoding time saving which is about 54% on average with only 0.7% BD-rate increase. In the future, average gradients along more directions can be exploited to obtain the candidate list with fewer modes for RMD and RDO. More effective features such as the intra mode and the variance of the coding block can be also considered in SVM for CU size decision to further improve the prediction accuracy."

Explore the future work suggested here and comment on further improvements in fast intra mode and CU size decision for HEVC. Extend the performance comparison similar to Figures 12, 14 and 15 and Tables III thru V.

P.5.235 See [E369] In this paper Liu et al. have proposed an adaptive and efficient mode decision algorithm based on texture complexity and direction for intra HEVC prediction. They have also presented a detailed review of previous work in this area. By developing the concepts of CU size selection based on texture complexity and prediction mode decision based on texture direction they have shown significant reduction in encoder complexity with negligible reduction in BD-rate using standard test sequences. Go through the texture complexity analysis and texture direction analysis described in this paper and confirm the comparison results shown in Tables V–X. Extend this comparison to 4K and 8K test sequences.

P.5.236 See [E270, E271] Using the early termination for TZSearch in HEVC motion estimation, evaluate the reduction in HEVC encoding time based on HDTV, UHDTV (4K and 8K) test sequences for different profiles.

P.5.237 S.K. Rao [E372] has compared the performance of HEVC intra with JPEG, JPEG 2000, JPEG XR, JPEG LS and VP9 intra using SD, HD, UHD and 4K test sequences. HEVC outperforms the other standards in terms of PSNR and SSIM at the cost of increased implementation complexity. Consider BD-PSNR and BD bit rate [E81, E82, E96, E198] as comparison metrics.

P.5.238 See P.5.237. Extend the performance comparison to 8K test sequences.

P.5.239 See [E373] The impact of SAO in loop filter in HEVC is shown clearly by Jagadeesh [E373] as part of a project. Extend this project to HDTV, UHDTV, 4K, and 8K video sequences.

P.5.240 See [E374] Following the technique proposed by Jiang and Jeong (see reference 34 cited in [E374]) for fast intra coding, Thakur has shown that computational complexity can be reduced by 14–30% compared to HM 16.9 with negligible loss in PSNR (only slight increase in bit rate) using Kimono (1920 × 1080), BQ Mall (832 × 488) and Kristen and Sara (1280 × 1080) test sequences. Extend this technique to 4K and 8K test sequences and evaluate the complexity reduction.

P.5.241 See [E375] following reference 2 cited in this paper, Sheelvant analyzed the performance and computational complexity of high efficiency video encoders under various configuration tools (see Table 2 for description of these tools) and has recommended the configuration settings (see conclusions). Go through this project in detail and see if you agree with these settings. Otherwise develop your own settings.

P.5.242 See [E357] Hsu and Shen have developed VLSI architecture of a highly efficient deblocking filter for HEVC. This filter can achieve 60 fps for the 4K×@k video under an operating frequency of 100 MHz. This design can achieve very high processing throughput with reduced or comparable area complexity. Extend this deblocking filter HEVC architecture for 8K × 4K video sequences.

P.5.243 See [E377] Bae, Kim and Kim have developed a DCT-based LDDP distortion model and then proposed a HEVC-complaint PVC scheme. Using various test sequences (see V–VII), they showed the superiority of their PVC approach compared with the state-of-the-art PVC scheme for LD and RA main profiles. Extend this technique to AI main profile.

P.5.244 See P.5.243 Extend this technique to 4K and 8K test sequences and develop Tables similar to those shown in [E377]. Based on these simulations, summarize the conclusions.

P.5.245 See [E379] Pagala's thesis focusses on multiplex/demultiplex of HEVC Main Profile video with HE-AAC v2 audio with lip synch. Extend this to 8K video test sequences. Give a demo.

P.5.246 See [E379]. In the multiplex/demultiplex/lip synch scheme, replace HEVC Main Profile with VP10 video. Make sure that the time lag between video and audio is below 40 ms. See [VP5, VP15]. Give a demo.

P.5.247 See [E379], P.5.245 and P.5.246. Implement multiplex/demultiplex/lip sync scheme using AV1 video and HE AAC v2 audio.

AV1 codec is developed by alliance for open media (AOM). http://tinyurl.com/zgwdo59

P.5.248 See [E379]. Replace AV1 video in P.247 by DAALA codec. See the references on DAALA.

P.5.249 See [E379]. Can this scheme be extended to SCC?

P.5.250 See [E382] This paper explains clearly the fully pipelined architecture for intra prediction in HEVC and achieves a high throughput of 4 pels per clock cycle. It can decode 3840 × 2160 videos at 30 fps. In the conclusions the authors state "in the future work, we plan to implement the proposed architecture on ASIC platform to increase the system frequency, aiming at achieving real-time video decoding of higher resolution (higher than 4K videos)". Explore this.

P.5.251 See P.5.250. The authors also state "Meanwhile the proposed architecture will be integrated with inter prediction engine, inverse transforming engine and other engines to construct the entire encoding/decoding system." This is a major project. Implement this HEVC codec.

P.5.252 See [E281]. In the conclusions it is stated "Then, two flexible and HEVC compliant architectures, able to support the DCT of size 4, 8, 16, 32 have been proposed". In beyond HEVC, DCT of size 64 also being considered. Extend these HEVC compliant architectures to support the DCT of size 64. See the sections beyond HEVC and projects on beyond HEVC toward the end.

P.5.253 Based on a set of Pareto-efficient encoding configurations identified through a rate-distortion-complexity analysis Correa et al. [E383] have developed a scheme that accurately limits the HEVC encoding time below a predefined target for each GOP. The results also indicate negligible BD-rate loss at significant complexity reduction. Go through this paper in detail and confirm the results shown in Tables II–IV and Figures 1 and 4–9 using the test sequences listed in Table I.

P.5.254 See [E384]. After describing the mosaic videos with arbitrary color filter arrays (Figure 1), the authors propose a novel Chroma subsampling strategy (4:2:0 format) for compressing mosaic videos in H.264]/AVC and HEVC. They claim that this strategy has the best quality and bitrate compared with previous schemes.

For the seven typical RGB-CFA structures shown in Figure 1 (captured mosaic videos) apply the demosaicking designed for each structure and as well the universal demosaicking and obtain the demosaicked full-color RGB images.

P.5.255 See P.5.254. For the demosaicked videos convert from RGB to YUV and then to 4:2:0 six subsampling strategies (Figure 6). Convert from 4:2:0 to 4:4:4 YUV format and then to RGB reconstructed video. Compare these strategies in terms of CPSNR and CMSE for these demosaicked videos.

P.5.256 Implement the chroma subsampling strategy described in [E384] along with other strategies (Figures 2–4) in both H.264/AVC and HEVC and confirm the results shown in Figures 5–8 and Tables I–VI.

P.5.257 Apply the universal demosaicking algorithm based on statistical MAP estimation developed in [E387] to the seven typical RGB-CFA structures (see P.5.284) and obtain the demosaicked full color RGB images. Compare the effectiveness of this universal algorithm with those described in [E384] in terms of the PSNR. See the conclusions section in [E387].

P.5.258 See [E388]. Also the abstract. Implement the fast prediction mode decision in HEVC developed in [E388] and confirm the simulation results shown in Tables 1–3.

P.5.259 See P.5.258. Apply this technique to 8K test sequences and develop tables similar to Tables 1-3 described in [E388].

P.5.260 Review [E389] in detail. The last sentence in Conclusion section states, "The future work will focus on extending algorithm to the remaining coding structures (i.e., PUs and TUs) and other configurations in order to further expedite the encoding process with minimal impact on the coding efficiency." Explore this and develop tables similar to Tables V–VII and draw the conclusions.

P.5.261 See [AVS12]. The authors proposed a fast intra coding platform for AVS2 (called iAVS2) leading to higher speeds and better balance between speed and compression efficiency loss especially for large size videos. The authors state "Owing to their (AVS2 and HEVC) similar frameworks, the proposed systematic solution and the fast algorithms can also be applied in HEVC intra coding design". Go through this paper in detail and explore how the speed up methods can be applied to intra HEVC using various test sequences based on the standard performance metrics.

P.5.262 See [E392]. SDCT and its integer approximation have been proposed and applied in image coding. The authors suggest the possibility to implement efficiently an integer SDCT in the HEVC standard.

Integrating inside HEVC may require a significant amount of work, as the transform has to be inserted in the rate-distortion optimization loop, and auxiliary information may have to be signaled. RD optimization is feasible but, the way at least the HEVC software is written, it is not necessarily easy. What can be easily done is to take the HEVC integer transform and rotate the transform using the technique developed in [E392] for obtaining a rotated transform. Implement this in all profiles and compare with the HEVC HM software. Consider both options.

P.5.263 See P.5.262 Using SDCT and its integer approximation in HEVC invariably results in increased implementation complexity. Investigate this thoroughly.

P.5.264 Performance comparison of SDCT with DCT is shown in Tables I, II, and IV [E392]. However the block sizes are limited to 32×32. For super high resolution videos it is suggested that even larger block sizes such as 64×64 are suggested. Extend these Tables for 64×64 block size.

P.5.265 See P.5.265. Extend these Tables to 4K and 8K video sequences.

P.5.266 Performance comparison of INTSDCT with DCT is shown in Table V [E392]. Extend this comparison to 4K and 8K video sequences and also larger block size such as 64×64.

P.5.267 In P.5.264 thru P.5.266 consider implementation complexity as another comparison metric.

P.5.268 Huang et al. [E393] have developed false contour detection and removal (FCDR) method and applied it to HEVC and H.264 videos as post processing operation. They also state "It will be interesting to adopt it as part of the in loop decoding process (Figure 5.5). This idea demands further investigation and verification". Investigate this in detail and see how FCDR can be embedded as an in loop operation besides the deblocking and SAO filters. Assuming FCDR embedding is successful, compare this with FCDR as a post processing operation in terms of PSNR, SSIM and subjective quality (See the corresponding figures and tables in [E393].) using various test sequences.

P.5.269 See P.5.268. Consider implementation complexity as another comparison metric in the FCDR process (in loop vs. post processing) in both HEVC and H.264.

P.5.270 Chen et al. [394] proposed a novel block-composed background reference (BCBR) scheme and is implemented in HEVC. They claim that the new BCBR algorithm can achieve better performance than baseline HEVC. This technique is however limited to sequences captured by static cameras (surveillance and conference sequences). They also suggest, "For moving camera cases, the long-term temporal correlation due to background is also worthy of investigation and the block-composed sprite coding would be a good choice". Investigate this in detail. Consider also encoding and decoding complexity as another performance metric.

P.5.271 See P.5.270. The authors in [394] conclude "We would like to extend our BCBR to more generic cases in our future work." Explore this.

P.5.272 Min, Xu and Cheung [E397] proposed a fully pipeline architecture, which achieves higher throughput, smaller area and less memory, for intra prediction of HEVC. They conclude "In the future work, we plan to implement the proposed architecture on ASIC platform to increase the system frequency, aiming at achieving real time video decoding of higher resolution.". Implement this.

P.5.273 See P.5.272 Min, Xu and Cheung [E397] further state "Meanwhile, the proposed architecture will be integrated with inter prediction engine, inverse transforming engine, and the other engines to construct the entire full scale encoding/decoding system.". Explore this in detail and implement the same.

P.5. 274 In [H.78] the authors state "We would like to extend this sketch attack framework to handle different video coding standards such as High Efficiency Video Coding (HEVC), Audio Video Standard (AVS) and Google VP9 as our future work.". Extend this attack to HEVC Main Profile.

P.5.275 In implementing the image codec based on SDCT, Francastoro, Fosson and Magli [E392] fixed the 8 quantization levels for the angles distributed uniformly between 0 and π. They state "In order to improve the compression performance, as future work, we may consider a non-uniform angle quantization." Investigate this thoroughly. Consider all possible non-uniform angle quantizations, the main objective being improved compression performance.

P.5.276 Wang et al. [E399] have described the MPEG internet video coding (IVC) standard and compared its performance (both objective and subjective – see Figures 10–13 for RA and LD) with web video coding (WVC), video coding for browsers (VCB) and AVC high profile. They have also compared the coding tools of IVC with those of AVC HP (Table IV). Using the test sequences and their configurations (Tables V–VII) confirm the results shown in Table VIII and Figures 10–13. This project requires detailed and extensive simulations and intensive viewing by naïve and expert subjects using the double stimulus impairment scale (DSIS).

P.5.277 See P.5.276. Simulations are based on classes A, B, and D test sequences (See Table V in [E399]). Extend these simulations to classes C, E, and F test sequences and draw the conclusions based on the IVC, WVC, VCB and AVC HP. As in P.5.276 this project also requires detailed and extensive simulations and intensive viewing by naïve and expert subjects using the double stimulus impairment scale (DSIS).

P.5.278 In the conclusion section Wang et al. [E399] state "Our team has set up an open source project xIVC, which aims to develop a real time IVC codec for HD sequences. As of now, the decoder of xIVC can decode 1080P sequences in real-time on PC platform." Develop this decoder.

P.5.279 See P.5.277. Develop the real time IVC codec (encoder/decoder) for HD sequences.

P.5.280 Xu et al. [E400] have developed video saliency detection using features in HEVC compressed domain and claim that their method superior to other state-of-the-art saliency detection techniques. They also state that their method is more practicable compared to uncompressed domain methods as both time and storage complexity on decoding videos can be saved. In the conclusions and future work the author's state

There exist three directions for the future work. 1) Our work in its present form merely concentrates on the bottom up model to predict video saliency. In fact, videos usually contain some top-down cues indicating salient regions, such as human faces. Indeed, an ideal vision system, like the one of humans, requires the information flow in both directions of bottom–up and top–down. Hence, the protocol, integrating the top-down model into our bottom-up saliency detection method, shows a promising trend in future. Investigate this in detail.

P.5.281 See P.5.280. "Many advanced tracking filters (e.g., Kalman filter and particle filter) have emerged during the past few decades. It is quite an interesting future work to incorporate our method with those filters, rather than the forward smoothing filter of this paper. In that case, the performance of our method may be further improved." Replace the forward smoothing filter adopted in [E400] by Kalman filter and particle filter and evaluate how the performance of the video saliency detection can be further improved.

P.5.282 See P.5.280. "A simple SVM learning algorithm, the C-SVC, was developed in our work for video saliency detection. Other state-of-the-art machine learning techniques may be applied to improve the accuracy of saliency detection, and it can be seen as another promising future work." Explore this future work.

P.5.283 See [E355]. In conclusion, the authors state "In the future more study can be made on how to facilitate software and hardware implementation." Implement tis.

P.5.284 See P.5.283. The authors [E355] further state "We also leave it as a further study how to efficiently apply CCP to various applications including high dynamic range video coding." Explore this for HDR video coding.

P.5.285 See [E355] Following the range extensions in HEVC (see the references at thee end), the authors have demonstrated, by using the CCP, significant coding performance improvements for both natural and screen content video. Verify this by confirming the experimental results shown in Table I–V and Figures 3 and 4.

P.5.286 See [E401] Section V Conclusion is repeated here:

In conclusion, it is evident that encoder could generate decoder resource optimized video bit streams by exploiting the diversities of the decoder complexity requirements of the HEVC coding modes. In this context, the proposed complexity model for HEVC inter-frame decoding predicts the decoding complexity with an average prediction error less than 5% for both uni- and bi-predicted frames. Furthermore, the proposed encoding algorithm is capable of generating HEVC bit streams that can achieve an average decoder complexity n reduction of 28.06 and 41.19% with a BD-PSNR loss of –1.91 and –2.46 dB for low-delay P and random access configurations, respectively, compared to the bit streams generated by the HM reference encoder. The future work will focus on extending the framework to consider both the rate and the distortion, alongside the decoder's complexity, to generate more optimized HEVC video bit streams.

Explore this future work and see how you can optimize this.

P.5.287 See [E402]. Section V Conclusion is repeated here:
In this work, an unthreaded version of the OpenHEVC decoder supporting parallel decoding at slice level is presented. Based on this decoder, implementations for several multicore chips, with and without threads support, have been tested with good results in both, performance and speedup. In the future, parallel decoding at frame level will be integrated, which certainly will improve the speedups with more than 4 cores
Integrate the parallel decoding Cabaratat the frame level and evaluate how the HEVC decoder can be speeded up.

P.5.288 A novel perceptual adaptive quantization method (AVQ) based on an adaptive perceptual CU splitting that improves the subjective quality of HEVC encoder along with less encoding time is proposed in [E403]. Confirm the simulation results shown in Table I and Figures 3 and 4.

P.5.289 See P.5.288. In Section 5 Conclusion, the authors [E403] state, "In the future, we will take the both the spatial and temporal characteristics into consideration to improve coding performance further." Explore this.

P.5.290 In Section V, the authors [E404] state, "The performance of proposed algorithm is more noticeable at high packet error rate and low bit rate conditions. Therefore, the proposed algorithm is suitable for video transmission applications in error-prone wireless channels which have bandwidth constraints. The perceived video quality can be improved by applying more sophisticated error concealment technique at the decoder.' Apply error concealment techniques at the HEVC decoder and investigate how the subjective video quality can be improved.

P.5.291 Boyadjis et al. [E405] have proposed a novel method for selective encryption of H.264/AVC (CABAC) and HEVC compressed bit streams. And compared with other encryption schemes (see the extensive list of references at the end). See the last para in the conclusion and future work of his paper. Explore these topics (studies).

P.5.292 Cabarat, Hamidouche, and Deforges [E410] have implemented a software parallel hybrid SHVC decoder (AVC decoder for BL and HEVC decoder for EL) (see Figure 2). The software design enables a real time decoding of the HEVC EL at 2160p60 while the AVC base layer is decoded at 1080p60 for x2 spatial scalability. Implement this SHVC software and confirm the results listed in Tables 1–4. The software sites listed in the references will be very helpful.

P.5.293 See P.5.292 The authors [E410] state that only RA configuration is used for hybrid SHVC decoder. Extend their research to AI and LD configurations. Based on the simulation results, what conclusions can be made?

P.5.294 See [E405] and Section V Conclusion.

Implement the proposed extended selective encryption and reconstruct the encrypted images shown in Figures 6–8. Repeat this scheme using other test sequences.

P.5.295 See P.5.294

In Section V Conclusion, the authors state
"We did not provide in this paper an exhaustive analysis of CABAC regular mode cipherable syntax elements. Some of them are easily identifiable (for example, the process proposed in this paper could almost identically apply to the intra-prediction modes for Chroma components), but a large majority of these cipherable elements would require a very specific monitoring, which we did not address in our study." Carry out this exhaustive analysis.

P.5. 296 See P.5.295 and P.5.296.

In Section V Conclusion, the authors further state
"An analysis of the *maximum* ES for both H.264/AVC and HEVC, based on the proposed *regular* mode encryption, is thus one major track for future research. Nevertheless, such a study would have to pay particular attention to a critical tradeoff we highlighted in our study and the underlying correlation between the improvement in the scrambling performance and its consequences on the compression efficiency."

Carry out this further research.

P.5.297 Oliveria et al. [DCT34] have implemented video coding based on low-complexity approximation of DTT and compared with H.264/AVC. Replace the integer DCT in HEVC [E61] by the integer DTT and compare its performance with the standard HEVC in video coding. Develop graphs similar to Figures 1 and 2 in [E62]. See DCT-P28 thru DCT-P34.

P.5.298 See P.5.297 The rate distortion graphs shown in Figures 1 and 2 in [E62] are for 1280x720 60 HZ and 1920x1080, 24 Hz video test sequences and class E and class B (see Table VIII in [E62]). Extend the performance comparison (integer DTT and integer DCT) based on video test sequences – classes A, C, D and E' listed in Table VIII.

P.5.299 Lainema et al. [E78] provided an overview of intra coding techniques in HEVC and compared its performance with H.264/AVC intra coding. Replace INTDCT by INTDTT [DCT34] in HEVC intra coding and compare its performance with H.264/AVC intra coding. Develop tables similar to those in the work done by Lainema et al. [E78] using the test sequences (Table IX).

P.5.300 The encoder block diagram for screen content coding is shown in Figure 8.3. Replace the blocks "Transform & Quant" and "De-quant & Inv. Transform" by the blocks "DTT & Quant" and "De-quant and Inverse DTT" respectively. Implement the SCC coding with the DTT/Inverse DTT and compare the results shown in [6] of Chapter 8 Screen Content Coding. Based on these results what conclusions can be drawn. This requires thorough understanding of screen content coding extension of HEVC as described in the review paper [6] of Chapter 8. Details on DTT are described in [E62].

P.5.301 Atta and Ghanbari [H79] developed an efficient rate control scheme for a joint temporal-quality scalability in H.264/AVC. They showed that this algorithm achieves better coding efficiency with low computational complexity compared to earlier ones. Explore how this rate control scheme can be adopted for a joint temporal-quality scalability in HEVC/H.265. See the overview paper [OP6] on scalable coding in HEVC (SHVC).

P.5.302 The performance of HEVC is compared [E62] with earlier standards such as H.262/MPEG-2, MP video, H.263 HLP, MPEG-4 ASP visual, and H.264/MPEG-4 AVC HP based on PSNR and subjective tests. The used test sequences are listed in Table VIII [E62]. Confirm the objective test results by accessing the software from the ftp site listed at the end just before the appendix in [E62].

P.5.303 See P.5.302. Compare the performance of HEVC with AVS China.

P.5.304 See P.5.302. Compare the performance of HEVC with DIRAC (BBC).

P.5.305 See P.5.302. Compare the performance of HEVC with DIRAC (BBC).

P.5.306 See P.5.302. Compare the performance of HEVC with THOR video codec.

P.5.307 See P.5.302. Compare the performance of HEVC with WMV9 (VC1).

P.5.308 See P.5.302. Compare the performance of HEVC with VP9.

P.5.309 See P.5.302. Compare the performance of HEVC with DAALA.

P.5.310 See P.5.302. Compare the performance of HEVC with AV1 codec (AOM).

P.5.311 See [E413] and the conclusions. By incorporating separately three parts 1) AIP with SDT-SVD, 2) TMP with SDT-SVD and 3) AIP with SDT-SVD+TMP (see Figure 5) the authors have showed some improvements in HEVC intra coding performance. However the encoding/decoding complexities are much higher (See Section C: computational complexity).
They state "In the future, fast algorithms, such as early skip for SDT-SVD transform and fast template searching method, can be exploited to reduce the complexity of the proposed method." Explore this.

P.5.312 See P.5.311. The authors further state "To further improve the efficiency of the SDT-SVD transform, more work can be done to design an efficient block with more similar structures as the residual block for deriving SVD transforms." Explore this.

P.5.313 See [E414]. The authors have integrated several techniques including AC energy of the CU that have resulted in significant reduction in encoder complexity for all intra mode (up to 71% reduction) with negligible increase in bit rate at nearly the same rate distortion. Review this paper thoroughly and confirm the simulation results shown in Figure 3.

8

Screen Content Coding for HEVC

Abstract

Screen content (SC) video is generated by computer programs and displayed on screen without any signal noise. The SC picture usually contains a lot of discontinuous textures and edges. Patch of SC picture is often identical in many regions in the picture. Based on these differences from camera-captured video, it has been developed as an extension of HEVC [6]. We discuss, here, the four screen content coding tools, lossless and lossy coding performance, fast algorithms, quality assessment, and other algorithms developed recently.

Keywords

Adaptive color transform, Adaptive motion vector resolution, Intra block copy, Palette coding, Residual DPCM, Sample-based prediction, Screen contents, Screen image quality assessment, String matching, Template matching, Transform skip

8.1 Introduction to SCC

High efficiency video coding (HEVC) discussed in Chapter 5 is the latest video compression standard of the Joint Collaborative Team on Video Coding (JCT-VC), which was established by the ITU-T Video Coding Experts Group (VCEG) and the ISO/IEC Moving Picture Experts Group (MPEG) [1]. It demonstrates a substantial bit-rate reduction over the existing H.264/AVC standard [2]. Several extensions and profiles of HEVC have been developed according to application areas and objects to be coded. However, both the HEVC and the H.264/AVC focused on compressing camera-captured video sequences, mainly consisting of human objects and complex textures (Figure 8.1). Although several different types of test sequences were used during the development of these standards, the camera-captured sequences exhibited common characteristics such as the presence of sensor noise and an abundance of translational motion. Thus, video compression exploits both temporal and spatial redundancies. A frame which is compressed by exploiting the spatial redundancies is termed as intra frame and the frames which are compressed by exploiting the temporal redundancies are termed as inter frames. The compression of an inter frame requires a reference frame which will be used to exploit the temporal redundancies. The inter frame is usually of two types namely a P frame and a B frame. The P frame make use of one already encoded/decoded frame which may appear before or after the current picture in the display order i.e. a past or a future frame as its reference, whereas the B frame makes use of two already encoded/decoded frames, one of which is a past and the other being the future frame as its reference frames, thus, providing higher compression but also higher encoding time as it has to use a future frame for encoding [3]. Furthermore, conventional video coders, in general, remove high-frequency components for compression purposes, since the human visual sensitivity is not so high in high frequencies.

Figure 8.1 Image example from camera captured video content [4].

Recently, however, there has been a proliferation of applications that display more than just camera-captured content. These applications include displays that combine camera-captured and computer graphics, wireless displays, tablets as second display, control rooms with high resolution display wall, digital operating room (DiOR), virtual desktop infrastructure (VDI), screen/desktop sharing and collaboration, cloud computing and gaming, factory automation display, supervisory control and data acquisition (SCADA) display, automotive/navigation display, PC over IP (PCoIP), ultra-thin client, remote sensing, etc. [5, 6]. The type of video content used in these applications can contain a significant amount of stationary or moving computer graphics and text, along with camera-captured content, as shown in Figure 8.2. However, unlike camera-captured content, screen content frequently contains no sensor noise, and such content may have large uniformly flat areas, repeated patterns, highly saturated or a limited number of different colors, and numerically identical blocks or regions among a sequence of pictures. These characteristics, if properly managed, can offer opportunities for significant improvements in compression efficiency over a coding system designed primarily for camera-captured natural content. Unlike natural images/video, screen contents may not be very smooth. They usually have totally different statistics. For text or graphics contents, it is much sharper and with high contrast [7]. Because of the high contrast, any little artifact caused by removing high frequency components in conventional video coders may be perceived by users. Thus, all coding techniques supported by HEVC RExt and additional coding tools such as intra block coping, palette coding, and adaptive color space transform are required to compress screen content. Features of screen contents are summarized as:

- Sharp content: Screen content usually includes sharp edges, such as in graphic or animation content. To help encoding sharp content, transform skip has been designed for screen content [8].
- Large motion: For example, when browsing a web page, a large motion exists when scrolling the page. Thus, new motion estimation algorithms to handle the large motions for screen content may be required.
- Artificial motion: For example, when fading in or fading out, the conventional motion model may not be easy to handle it.
- Repeating patterns: For example, the compound images may contain the same letter or objects many times. To utilize the correlation among repeating patterns, Intra Block Copy (IBC) [6, 20, 43, 69, 72] has been developed.

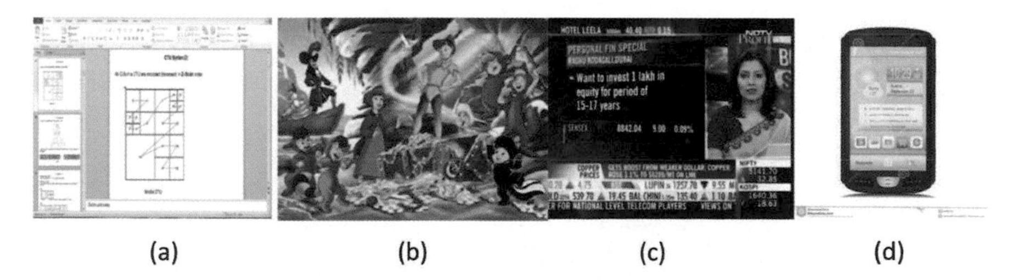

Figure 8.2 Images of screen content: (a) slide editing, (b) animation, (c) video with text overlay, (d) mobile display [4].

Joint Call for Proposals (CfP) was released in Jan. 2014 with the target of developing extensions of the HEVC standard including specific tools for screen content coding [9]. The use cases and requirements of the CfP are described in [5] and common conditions for the proposals are found in [10]. These documents identified three types of screen content: mixed content, text and graphics with motion, and animation. Up to visually lossless coding performance was requested for RGB and YC_bC_r 4:4:4 format having 8 or 10 bits per color component. After seven responses to the CfP were evaluated at the JCT-VC meeting [11], several core experiments (CEs) were defined including intra block copying extensions, line-based intra copy, palette mode, string matching for sample coding, and cross-component prediction and adaptive color transforms. As a result of evaluating the outcome of the CEs and related proposals, the HEVC Screen Content Coding Test Model 6 [12] and Draft Text 5 [13] were published in Oct. 2015. All the documents are available in [14]. Test sequences [15], reference software [16] and manual [17] are also available.

8.2 Screen Content Coding Tools

HEVC-SCC is based on the HEVC framework while several new modules/ tools are added as shown in Figure 8.3, including intra block copy (IBC) [72], palette coding, adaptive color transform, and adaptive motion vector resolution.

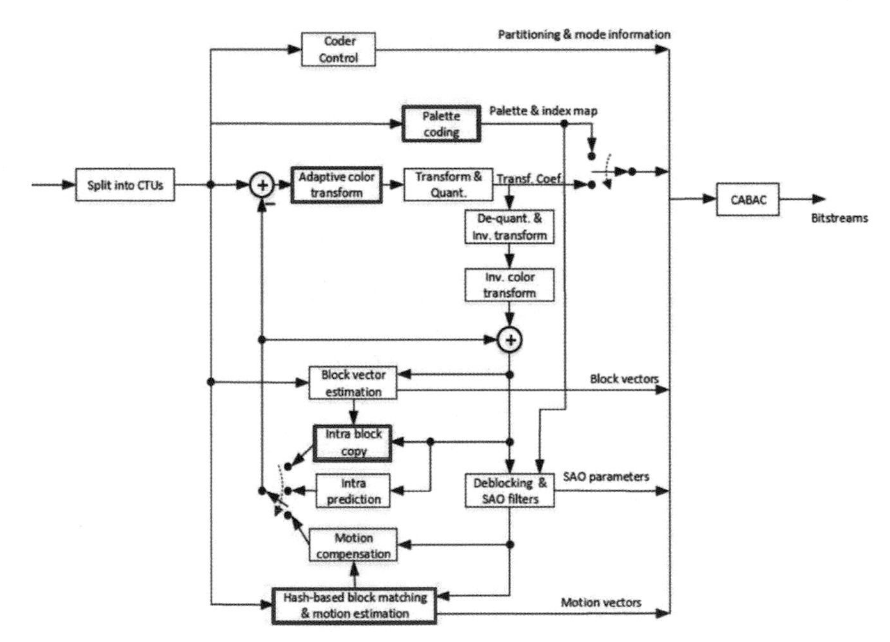

Figure 8.3 Encoding block diagram for screen content coding [6] © IEEE 2016.

8.2.1 Intra Block Copy

HEVC-SCC introduces a new CU mode in addition to the conventional intra and inter modes, referred to as intra block copy (IBC). The IBC mode performs like an inter mode prediction but the prediction units (PU) of IBC coded coding units (CU) predict reconstructed blocks in the same picture, taking the advantage of exploiting the repeated patterns that may appear in screen content. Similar to inter mode, IBC uses block vectors to locate the predictor block [12].

Since screen content is likely to have similar or repeated patterns on a screen, such spatial redundancy can be removed by quite different way from the conventional intra prediction schemes. The most significant difference is the distance and shapes from neighboring objects [18]. Removable spatial redundancy in the conventional intra prediction schemes refers to the similarity between the boundary pixels of the block to be coded and the adjacent pixels located spatially within one pixel. However, the removable spatial redundancy in IBC mode refers to the similarity between the area in the reconstructed picture and the block to be coded. A target 2D block/object is predicted from a reconstructed 2D block/object that is more than one pixel distant from it using the motion or location information (motion vector) from the reference block/object.

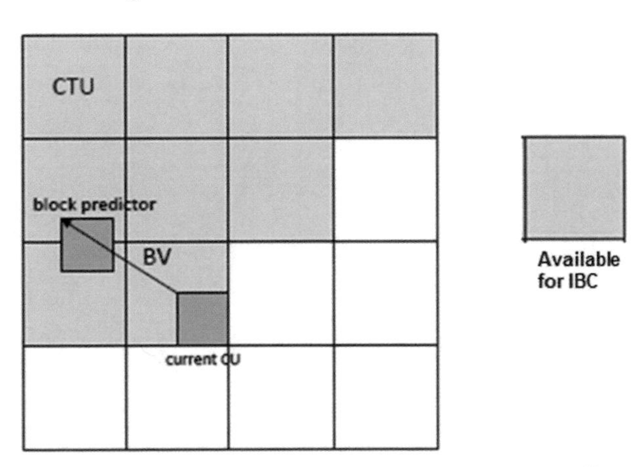

Figure 8.4 Intra block copy prediction in the current picture [6] © IEEE 2016.

Figure 8.4 shows the concept of IBC and the block vector (BV), which is conceptually similar to the motion vector (MV) of inter prediction. In terms of the accuracy of the vectors, MV has in usual quarter-pel accuracy to ensure improved prediction accuracy, whereas BV is enough to be in integer-pel accuracy. This is because of the characteristics of the screen content in IBC mode. For example, objects in computer graphics are generated pixel by pixel and repeated patterns are found in integer-pel accuracy. Block compensation, which is similar to motion compensation, of IBC is conducted on the reconstructed area of the current frame not previously coded, or decoded frames. In addition, the BV should be sent to the decoder side, but it is derived by prediction to reduce the amount of data in a manner similar to the motion vector. Prediction may be independent from the MV prediction method or in the same way as the MV prediction method.

The global BV search is performed for 8×8 and 16×16 blocks. Search area is a portion of the reconstructed current picture before loop filtering, as depicted in Figure 8.4. Additionally, when slices/tiles are used, the search area is further restricted to be within the current slice/tile. For 16×16 blocks, only a one-dimensional search is conducted over the entire picture. This means that the search is performed horizontally or vertically. For 8×8 blocks, a hash-based search is used to speed up the full picture search. The bit-length of the hash table entry is 16. Each node in the hash table records the position of each block vector candidate in the picture. With the hash table, only the block vector candidates having the same hash entry value as that of the current block are examined [19].

The hash entries for the current and reference blocks are calculated using the original pixel values. The 16-bit hash entry H is calculated as

$$\begin{aligned} H = \mathrm{MSB}(\mathrm{DC0}, 3) \ll 13 + \mathrm{MSB}(\mathrm{DC1}, 3) \ll 10 + \mathrm{MSB}(\mathrm{DC2}, 3) \\ \ll 7 + \mathrm{MSB}(\mathrm{DC3}, 3) \ll 4 + \mathrm{MSB}(\mathrm{Grad}, 4) \end{aligned} \quad (8.1)$$

where $\mathrm{MSB}(X, n)$ represents the n MSB of X, DC0, DC1, DC2, DC3 denote the DC values of the four 4×4 sub-blocks of the 8×8 block, and Grad denotes the gradient of the 8×8 block. Operator '$\ll$' represents arithmetic left shift.

In addition to full block vector search, some fast search and early termination methods are employed in the HEVC-SCC. The fast IBC search is performed after evaluating the RD cost of inter mode, if the residual of inter prediction is not zero. The SAD-based RD costs of using a set of block vector predictors are calculated. The set includes the five spatial neighboring block vectors as utilized in inter merge mode (as shown in Figure 8.5) and the last two coded block vectors. In addition, the derived block vectors of the blocks pointed to by each of the aforementioned block vector predictors are also included. This fast search is performed before the evaluation of intra prediction mode. It is applied only to 2N × 2N partition of various CU sizes.

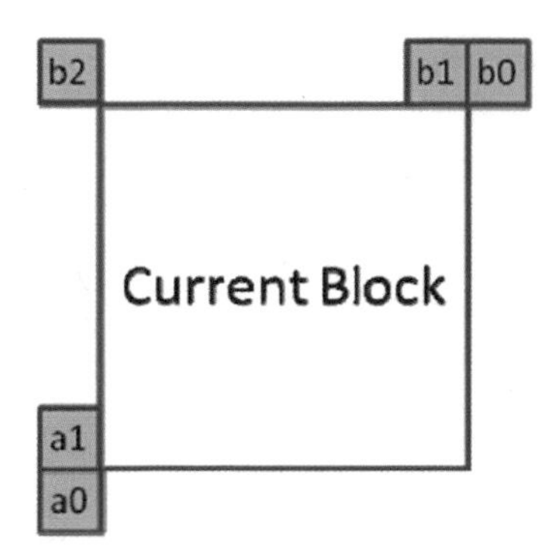

Figure 8.5 Candidates of block vector predictor [20] © IEEE 2015.

8.2.2 Palette Mode

For screen content, it is observed that for many blocks, a limited number of distinct color values may exist. In this case, the set of color values is referred to as the palette. Thus, palette mode enumerates those color values and then for each sample, sends an index to indicate to which color it belongs. In special cases it is also possible to indicate a sample that is outside the palette by signalling an escape symbol followed by component values as illustrated in Figure 8.6. Palette mode can improve coding efficiency when the prediction does not work due to low redundancy and when the number of pixel values for the block is small [18, 21]. According to the results by Xiu, et al. [22], coding gain of the palette-based coding increases up to 9.0% in the average BD-rate for lossy coding and up to 6.1% for lossless coding mode.

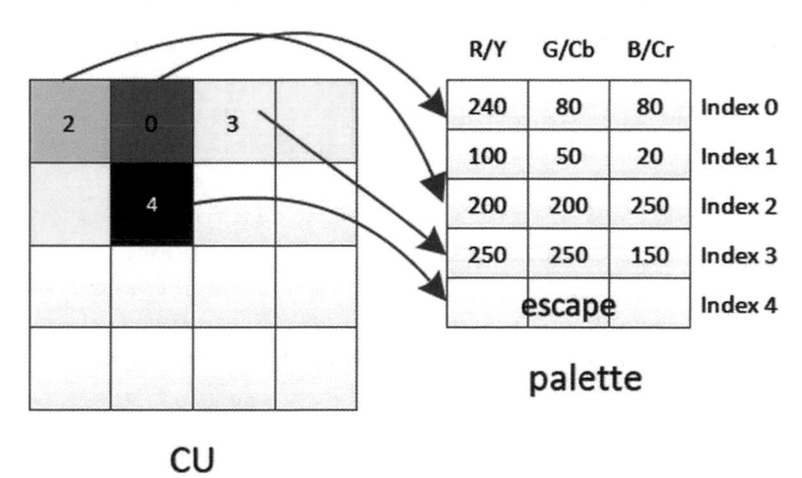

Figure 8.6 Example of indexed representation in the palette mode [6] © IEEE 2016.

8.2.2.1 Palette derivation

In the SCM-7.0 software [16], for the derivation of the palette for lossy coding, a modified k-means clustering algorithm is used. The first sample of the block is added to the palette. Then, for each subsequent sample from the block, the SAD from each of the current palette entries is calculated. If the distortion for each of the components is less than a threshold value for the palette entry corresponding to the minimum SAD, the sample is added to the cluster belonging to the palette entry. Otherwise, the sample is added as a new palette entry. When the number of samples mapped to a cluster exceeds a threshold, a centroid for that cluster is calculated and becomes the palette entry corresponding to that cluster.

In the next step, the clusters are sorted in a decreasing order of frequency. Then, the palette entry corresponding to each entry is updated. Normally, the cluster centroid is used as the palette entry. But a rate-distortion analysis is performed to analyze whether any entry from the palette predictor may be more suitable to be used as the updated palette entry instead of the centroid when the cost of coding the palette entries is taken into account. This process is continued till all the clusters are processed or the maximum palette size is reached. Finally, if a cluster has only a single sample and the corresponding palette entry is not in the palette predictor, the sample is converted to an escape symbol. Additionally, duplicate palette entries are removed and their clusters are merged.

For lossless coding, a different derivation process is used. A histogram of the samples in the CU is calculated. The histogram is sorted in a decreasing order of frequency. Then, starting with the most frequent histogram entry, each entry is added to the palette. Histogram entries that occur only once are converted to escape symbols if they are not a part of the palette predictor.

After palette derivation, each sample in the block is assigned the index of the nearest (in SAD) palette entry. Then, the samples are assigned to 'INDEX' or 'COPY_ABOVE' mode. For each sample for which either 'INDEX' or 'COPY_ABOVE' mode is possible, the run for each mode is determined. Then, the cost (in terms of average bits per sample position) of coding the mode, the run and possibly the index value (for 'INDEX' mode) is calculated. The mode for which the cost is lower is selected. The decision is greedy in the sense that future runs and their costs are not taken into account.

8.2.2.2 Coding the palette entries

For coding of the palette entries, a palette predictor is maintained. The maximum size of the palette as well as the palette predictor is signaled in the sequence parameter set (SPS). In SCM 4, a palette_predictor_initializer_present_flag is introduced in the PPS. When this flag is 1, entries for initializing the palette predictor are signaled in the bitstream. The palette predictor is initialized at the beginning of each CTU row, each slice and each tile. Depending on the value of the palette_predictor_initializer_present_flag, the palette predictor is reset to 0 or initialized using the palette predictor initializer entries signaled in the picture parameter set (PPS). In SCM 5, palette predictor initialization at the SPS level was introduced to save PPS bits when a number of PPS palette predictor initializers shared common entries. In SCM 6, a palette predictor initializer of size 0 was enabled to allow explicit disabling of the palette predictor initialization at the PPS level.

For each entry in the palette predictor, a reuse flag is signaled to indicate whether it is part of the current palette. This is illustrated in Figure 8.7. The reuse flags are sent using run-length coding of zeros. After this, the number of new palette entries is signaled using exponential Golomb code of order 0. Finally, the component values for the new palette entries are signaled.

previous palette

Index	G/Y	B/Cb	R/Cr
0	G0	B0	R0
1	G1	B1	R1
2	G2	B2	R2
3	G3	B3	R3
4	G4	B4	R4
5	G5	B5	R5

current palette

Pred flag	Index	G/Y	B/Cb	R/Cr
1	0	G0	B0	R0
0				
1	1	G2	B2	R2
1	2	G3	B3	R3
0				
0				
	3	G3N	B3N	R3N
	4	G4N	B4N	R4N

Re-used palette entries (3)

New palette entries (2), signalled

Figure 8.7 Use of palette predictor to signal palette entries [6] © IEEE 2016.

8.2.2.3 Coding the palette indices

The palette indices are coded using three main palette sample modes: INDEX mode, COPY_ABOVE mode, and ESCAPE mode as illustrated in Figure 8.8. In the INDEX mode, run-length coding is conducted to explicitly signal the color index value, and the mode index, color index, and run-length are coded. In the COPY_ABOVE mode, which copies the color index of the row above, the mode index and run-length are coded. Finally, in the ESCAPE mode, which uses the pixel value as it is, the mode index and the quantized pixel value are coded. When escape symbol is part of the run in 'INDEX' or 'COPY_ABOVE' mode, the escape component values are signalled for each escape symbol.

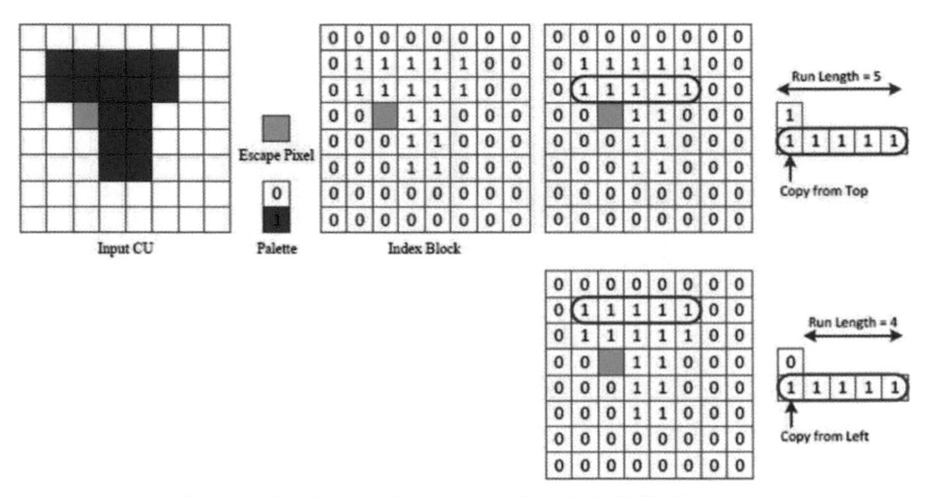

Figure 8.8 Coding the palette indices [21] © IEEE 2014.

8.2.3 Adaptive Color Transform (ACT)

Conventional natural content is usually captured in RGB color format. Since there is strong correlation among different color components, a color space conversion is required to remove inter-component redundancy. However, for screen content, there may exist many image blocks containing different features having very saturated colors, which leads to less correlation among color components. For those blocks, coding directly in the RGB color space may be more effective. ACT enables the adaptive selection of color-space conversion for each block. To keep the complexity as low as possible, the color-space conversion process is applied to the residual signal as shown in Figure 8.9 and after the intra- or inter-prediction process, the prediction residuals are selected to perform forward color-space transform as shown in Figure 8.10.

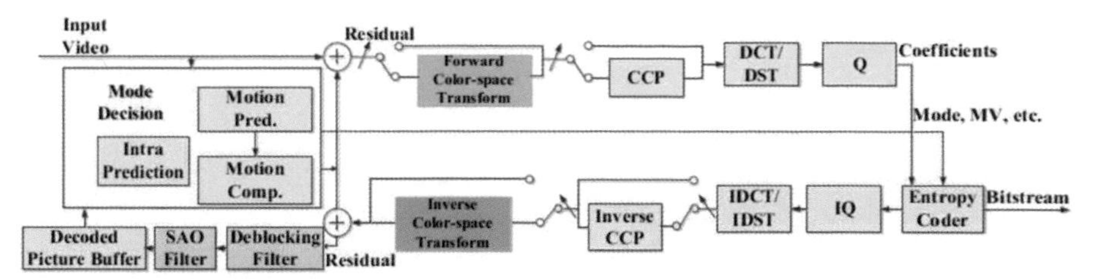

Figure 8.9 Location of the ACT in the encoder [23] © IEEE 2015.

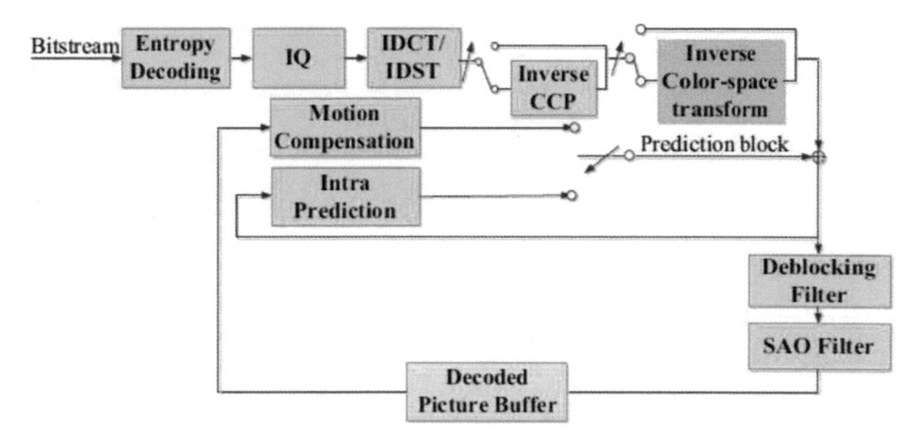

Figure 8.10 Location of the ACT in the decoder [23] © IEEE 2015.

8.2.3.1 Color space conversion

To handle different characteristics of image blocks in screen content, a RGB-to-YC_oC_g conversion [24] was investigated to use it for forward and backward lossy and lossless coding.

Forward transform for lossy coding (non-normative):

$$\begin{bmatrix} Y \\ Co \\ Cg \end{bmatrix} = \begin{bmatrix} 1 & 2 & 1 \\ 2 & 0 & -2 \\ -1 & 2 & -1 \end{bmatrix} \begin{bmatrix} R \\ G \\ B \end{bmatrix} /4 \tag{8.2}$$

Forward transform for lossless coding (non-normative):

$$\begin{aligned} Co &= R - B \\ t &= B + (Co \gg 1) \\ Cg &= (G - t) \\ Y &= t + (Cg \gg 1) \end{aligned} \tag{8.3}$$

Backward transform (normative):

$$
\begin{array}{l}
if(lossy)\{ \\
\quad Co = Co \ll 1 \\
\quad Cg = Cg \ll 1 \\
\} \\
\begin{array}{rcc}
t & = & Y - (Cg \gg 1) \\
G & = & Cg + t \\
B & = & t - (Co \gg 1) \\
R & = & Co + b
\end{array}
\end{array}
\tag{8.4}
$$

8.2.3.2 Encoder optimization

In the ACT mode, encoder complexity increases double because the mode searching is performed in both the original color space and the converted color space. To avoid this, following fast methods are applied:

- For intra coding mode, the best luma and chroma modes are decided once and shared between the two color spaces.
- For IBC and inter modes, block vector search or motion estimation is performed only once. The block vectors and motion vectors are shared between the two color spaces.

8.2.4 Adaptive Motion Vector Resolution

For natural video content, the motion vector of an object is not necessarily exactly aligned to the integer sample positions. Motion compensation is, therefore, not limited to using integer sample positions, i.e. fractional motion compensation is more efficient to increase compression ratio. Computer-generated screen content video, however, is often generated with knowledge of the sample positions, resulting in motion that is discrete or precisely aligned with sample positions in the picture. For this kind of video, integer motion vectors may be sufficient for representing the motion. Savings in bit-rate can be achieved by not signaling the fractional portion of the motion vectors.

Adaptive MV resolution allows the MVs of an entire picture to be signaled in either quarter-pel precision (same as HEVC version 1) or integer-pel precision. Hash based motion statistics are kept and checked in order to properly decide the appropriate MV resolution for the current picture without relying on multi-pass encoding. To decide the MV precision of one picture, blocks are classified into the following categories:

- C: number of blocks matching with collocated block
- S: number of blocks not matching with collocated block but belong to smooth region. For smooth region, it means every column has a single pixel value or every row has a single pixel value.
- M: number of blocks not belonging to C or S but can find a matching block by hash value.

The MV resolution is determined as:

- If CSMRate < 0.8, use quarter-pel MV.
- Otherwise, if C == T, use integer-pel MV.
- Otherwise, if AverageCSMRate < 0.95, use quarter-pel MV.
- Otherwise, if M $>$ (T-C-S)/3, use integer-pel MV.
- Otherwise, if CSMRate > 0.99 and MRate > 0.01, use integer-pel MV.
- Otherwise, if AverageCSMRate + AverageMRate > 1.01, use integer-pel MV.
- Otherwise, use quarter-pel MV.

T is the total number of blocks in one picture. CSMRate = (C+S+M)/T, MRate = M/T. AverageCSMRate is the average CSMRate of current picture and the previous 31 pictures. AverageMRate is the average MRage of the current picture and the previous 31 pictures.

8.3 Lossless and Visually Lossless Coding Algorithms

8.3.1 Residual DPCM

Differential pulse code modulation (DPCM) has been widely used to reduce spatial and temporal redundancy in the video content. The subtraction of the prediction signal from the current block signal generates the residual signal or the prediction error, containing the part of the original signal which could not be predicted by the selected predictor [25]. The residual signal can be further compressed by any method. In the HEVC, compression can be achieved by the application of a transformation, which is applied to represent the correlated parts of the residual signal in the residual block by a potentially small number of transform coefficients. These coefficients are then quantized and coded into the bitstream.

Sample-by-sample residual DPCM (RDPCM) of intra-predicted residuals was proposed [26] in the context of H.264/AVC lossless coding. When using this technique, instead of performing conventional intra-prediction each residual sample is predicted from neighboring residuals in the vertical or horizontal direction when the intra prediction is equal to one of these two directions. Average bitrate reductions of 12% were reported using this technique compared with conventional H.264/AVC lossless coding. This technique was later extended and adapted to the HEVC standard [27] achieving on average 8.4% bitrate reductions on screen content sequences.

The residuals of each sample are calculated by sample-by sample DPCM in vertical and/or horizontal mode. When the intra prediction mode is vertical, the RDPCM elements $\tilde{r}_{i,j}$ is given by

$$\tilde{r}_{i,j} = \begin{cases} r_{i,j} & , i = 0, 0 \le j \le (N-1) \\ r_{i,j} - r_{(i-1),j} & , 1 \le i \le (M-1), 0 \le j \le (N-1) \end{cases} \tag{8.5}$$

or when the intra prediction mode is horizontal mode

$$\tilde{r}_{i,j} = \begin{cases} r_{i,j} & , 0 \le i \le (M-1), j = 0 \\ r_{i,j} - r_{i,(j-1)} & , 0 \le i \le (M-1), 1 \le j \le (N-1) \end{cases}. \tag{8.6}$$

In the vertical mode, the samples in the first row in the block are left unchanged. All other samples are predicted from the sample immediately above in the same column. The horizontal intra RDPCM is given in a similar way.

The RDPCM elements are signaled to the decoder so that the original residual samples are reconstructed by

$$r_{i,j} = \sum_{k=0}^{i} \tilde{r}_{k,j} \quad , 0 \le i \le (M-1), 0 \le j \le (N-1), \text{in vertical mode}$$

$$r_{i,j} = \sum_{k=0}^{j} \tilde{r}_{i,k} \quad , 0 \le i \le (M-1), 0 \le j \le (N-1), \text{in horizontal mode.} \tag{8.7}$$

Thus, the RDPCM is implemented in one-dimensional direction in the HEVC-SCC. However, two dimensional RDPCM was suggested as [28]

$$\tilde{r}_{i,j} = \alpha_1 r_{i,j-1} + \alpha_2 r_{i-1,j} + \alpha_3 r_{i-1,j-1} + \alpha_4 r_{i-2,j} + \cdots \tag{8.8}$$

where α_i denotes weighting factor for neighboring pixels.

To find the best mode for the current residual block, the SAD distortion metric can be used for each mode (i.e. horizontal, vertical or no RDPCM mode). The mode with minimum SAD is selected as the best.

When performing the RDPCM on a block, samples in the first column and the first row for horizontal and vertical RDPCM, respectively, are not predicted. Therefore it is beneficial to exploit redundancy by performing prediction on these samples in the direction orthogonal to the main RDPCM direction, as shown in Figure 8.11.

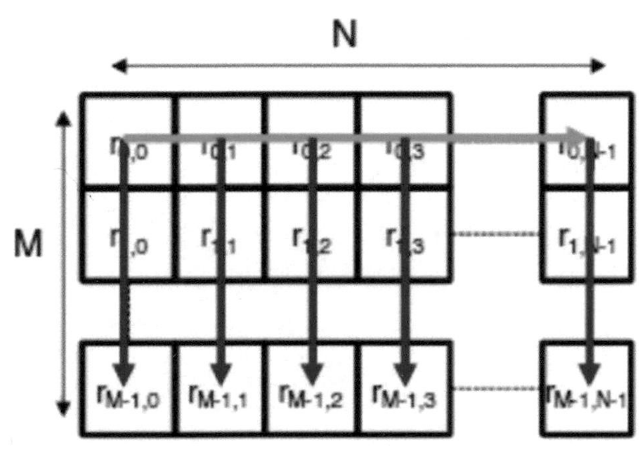

Figure 8.11 Secondary prediction (green line) after vertical RDPCM (red lines) [29] © IEEE 2013.

8.3.2 Sample-Based Weighted Prediction with Directional Template Matching

The sample-based weighted prediction (SWP) algorithm is proposed in [30] to introduce a weighted averaging of neighboring pixels for intra-prediction of the current pixel. The predicted pixel $p_{SWP}[i]$ is calculated as follows:

$$p_{SWP}[i] = \text{round}\left(\sum_{j \in S} w_{\text{int}}[i,j]g[j] / \sum_{j \in S} w_{\text{int}}[i,j]\right), \qquad (8.9)$$

where $g[j]$ is the pixels around the reconstructed current pixel, S is the set of supporting pixels, and $w_{\text{int}}[i,j]$ are the integer weighting values, which are calculated as follows:

$$w_{\text{int}}[i,j] = \text{round}\left(a_{SWP} \cdot b_{SWP}^{-SAD(P_g^k[i], P_g^k[j])/h_{dist}}\right), \qquad (8.10)$$

where the factor a_{SWP} is chosen to be 2^{11}, if the internal bit depth is 8, the basis factor b_{SWP} is 2 for exponential decaying weights, and the parameter

h_{dist} is empirically chosen to be 4.75 for luma and for 4:4:4 chroma. $SAD(\cdot)$ is the operator for the sum of differences between two patches, $P_g^k[i]$ and $P_g^k[j]$, which are formed by causally neighboring pixels (typically four pixels as shown in Figure 8.12). Thus, the $SAD(\cdot)$ can be defined as a similarity measure in the supporting area and given by

$$SAD\left(P_g^k[i], P_g^k[j]\right) = \sum_{v \in N_0} |g[i+v] - g[j+v]|\,. \tag{8.11}$$

Pixel X is predicted by a weighted average from the candidate pixels a, b, c, and d. For each candidate pixel, the SAD of the corresponding patches is calculated, e.g., for calculation of the SAD between X and b, the patch for X (pixels a, b, c, and d in the center of the figure) is compared to the patch for b (shaded blue area on the right hand side of the figure).

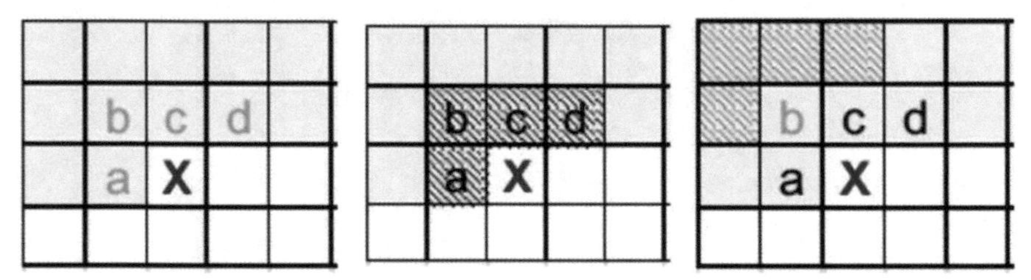

Figure 8.12 Causally neighboring pixels for prediction and patches. Pixel X is to be predicted from the patch with pixels a, b, c, and d (left, center); Patch around pixel b (right) [30] © IEEE 2013.

The prediction performance of the SWP is well for natural images that maintain high correlation among neighboring pixels, while degrades for such as text images that contain sharp edges among letters. With the weighted/averaged prediction, sharpness may be lost by averaging effect. Thus, another compromising prediction technique, directional template matching (DTM) has to be introduced. The main idea is to reduce the averaging effect by selecting the minimum SAD patch within the support area. The sharp edges could be kept without smoothing.

8.3.3 Sample-Based Angular Intra-Prediction

HEVC has adopted block-based angular intra-prediction which is useful for lossless coding to exploit spatial sample redundancy in intra coded CUs. A total of 33 angles (Figures 5.8 and 5.9) are defined for the angular prediction that can be categorized into three classes: 1 diagonal, 16 horizontal, and

16 vertical predictions. In HEVC, the total number of intra prediction modes is 35, including Mode 0 for INTRA_PLANAR, Mode 1 for INTRA_DC, and Mode 2 to 34 for INTRA_ANGULAR [31]. Given an $N \times N$ prediction unit (PU), the number of reference samples are $4N + 1$, i.e., 2N upper, 2N left, and 1 diagonal, belonged to neighboring PUs. All the samples inside the PU share the same prediction angle in the block-based angular intra-prediction. The value of prediction angle should be informed to the decoder. However, the sample-based angular prediction is performed sample by sample. Since four effective intra prediction block sizes ranging from 4×4 to 32×32 samples, each of which supports 33 distinct directions, there are 132 combinations of block sizes and prediction directions. The prediction accuracy is 1/32 in the horizontal or vertical direction via linear interpolation.

The coding gain of the SAP provides a 1.8% to 11.8% additional bitrate reduction on average [32] in the lossless coding mode. In addition the SAP provides more gain in 10-bit configurations due to the fact that there are more blocks with large prediction residuals, the difference between the original pixel value and its prediction, in the 10-bit video than in 8-bit video. The SAP also improves coding efficiency by increasing the usage of angular intra prediction and the number of intra coded CUs, while decreasing the usage of the planar and DC modes.

8.3.4 Sample-Based Angular Intra-Prediction with Edge Prediction

Another type of compound contents includes whole slide images (WSIs), which is the digitized version of microscope glass slides. Pathologists can send WSIs to others for sharing, collaborating, consulting and making diagnosis. WSIs usually feature a high number of edges and multifunctional patterns due to great variety of cellular structures and tissues. They should be scanned at high resolutions, resulting in huge file sizes. Therefore, designing efficient and accurate lossless or visually lossless compression algorithms are an important challenge. Edge prediction is proposed in [33] as a post-processing step on the residual signal computed by the original intra coding process. This method adds an extra coding step to the pipeline and alters the block-wise coding structure of HEVC as in [30].

In order to maintain the inherent block-wise coding and decoding structure of HEVC, an alternative intra coding modes are suggested in [34], while HEVC-RExt includes the optional use of the SAP, which is limited to the horizontal and vertical directions. For the case of the DC mode, a sample prediction is computed as an average of neighboring samples at positions $\{a, c\}$; $P_{x,y} = (a + c) \gg 1$ in Figure 8.12. For the case of the PLANAR mode, an edge predictor is proposed and calculated as follows:

$$P_{x,y} = \begin{cases} \min(a, c) & \text{if } c \geq \max(a, c) \\ \max(a, c) & \text{if } c \leq \min(a, c) \\ a + c - b & \text{otherwise} \end{cases} \tag{8.12}$$

The edge predictor mode and the SAP modes require that samples be decoded sequentially and be readily available for the prediction and reconstruction of subsequent samples. This inevitably breaks the block-wise decoding structure of HEVC. However, the reconstruction can be regarded as a spatial residual transform that only depends on the residual samples and the reference samples. Such spatial residual transform can be expressed in matrix form and applied during the decoding process in order to maintain the block-wise decoding structure. This technique improves coding efficiency by an average of 6.64% compared to SAP-1, which applies the SAP in all angular modes with a constant displacement among any two adjacent modes, and by an average of 7.67% compared to SAP-HV, which applies the SAP only in the pure horizontal and vertical directions [34]. This is mainly due to the introduction of a DC mode based on the SAP and an edge predictor in lieu of the PLANAR mode.

Although the edge predictor is capable of detecting horizontal and vertical edges accurately by selecting the best predictor for each pixel, more efficiency needs to be achieved in textual smooth regions by taking median values of adjacent five pixels as

$$P_{x,y} = \text{median}\{a, b, c, d, e\} \tag{8.13}$$

By adding the median prediction, coding efficiency increases by 16.13% on average compared to HEVC intra prediction coding [35].

8.4 Fast Coding Algorithms

As discussed in Chapter 5, the design of HEVC was based on two main objectives: achieving higher coding efficiency compared to previous standards and attaining low enough complexity to enable video applications on mobile devices. The second objective, however, is not easily fulfilled due to highly complex algorithms such as quadtree-based coding structure, large block transforms, an advanced motion prediction, additional filtering operations, and a high number of intra-prediction modes [36]. Among those properties, motion prediction and intra-prediction can be taken into account to design the screen content codec, responding to different nature from camera captured sequences.

8.4.1 Adaptive Motion Compensation Precision

Fractional precision motion compensation usually improves the video coding efficiency, especially for natural video. But considering the fact that slow moving or text-based screen content may be motion compensated in integer precision, using fractional precision is a waste of bits. Thus, adaptive precision methods are developed to improve coding efficiency of screen content, saving bits by using integer motion vectors in some cases. Encoder is designed to use integer precision when encoding the content captured from a screen and to use fractional precision when encoding the content captured from a normal camera.

To adopt the adaptive precision method in the HEVC structure, encoder should signal a flag at the slice header level to indicate whether integer precision or fractional precision is used for the current slice. Moreover, two-pass encoding is required to decide the precision of motion vectors, taking approximately double the encoding time, which is undesirable in practical use. Thus, fast algorithm has to be developed to minimize the encoding time while preserving most of the benefits brought by adaptive motion compensation precision.

In [37], fast algorithm is developed to efficiently design hash-based block matching scheme, using cyclic redundancy check (CRC) as the hash function to generate the hash value for every block. The complexity of CRC operation is about $O(m \cdot n \cdot w \cdot h)$, by considering the block size and picture size. More than 8G operations are required to calculate all the hash values of 64×64 blocks in 1080p video [38], for example. To reduce the complexity of hash table generation, reuse of the intermediate results is suggested. In fact, there exists $(m - 1)$ overlapping rows between two block operations. Thus, 63 of the intermediate data could be reused in the next 64×64 block, resulting in reduced complexity $O(m \cdot w \cdot h + n \cdot w \cdot h)$, i.e., about 256M operations for the same example. Next step is the block matching using the hash values and another reduction is possible. This can be done by checking all the blocks having the same hash value and selecting a block to predict the current block. The complexity of block matching is about $O(l \cdot (m \cdot n))$, where l denotes the number of blocks in a picture, resulting in about 2M operations for example above. The overall complexity of the proposed hash-based block matching method is about $O(m \cdot w \cdot h + n \cdot w \cdot h + l \cdot m \cdot n)$, resulting in reduced complexity from 4T (full picture search) to 258M, which is more than 15,000 × speedup.

Although we can reduce the complexity in the block matching operations, further reduction is possible using the adaptive precision decision for screen content. For example, in [37], the blocks are classified into four categories: collocated matched blocks (C) that the optimal motion vector should be (0, 0), smooth blocks (S) that every row or column has a single pixel value, matched blocks (M) that an exact match by hash values can be found, and other extra blocks (O). Some logical analysis can be given as follows:

- If the percentage of O is too large, fractional precision MV is used.
- If the percentage of M is not large enough, fractional precision MV is used.
- If the percentage of C, S, and M is larger than a threshold, integer precision MV is used.

Using the adaptive precision MV, the maximum bit saving was 7.7% for the YUV Desktop sequence under Low Delay coding structure without a significant impact on encoding time.

8.4.2 Fast Intra Coding

HEVC SCC has adopted several enhanced coding methods to improve compression efficiency by considering the properties of computer generated contents. Due to lot of redundancy in the spatial domain, intra prediction and processing are mainly dealt with in the standard. Slight changes can be made to generate coding tree unit (CTU) and other intra-based tools such as intra block copy, palette mode, adaptive color transform are newly introduced. Many researches are focused on these intra coding techniques that a large amount of computation is required.

CTU is a quad-tree structure that can be a CU or can be split into four smaller units, if necessary. Since the SCC often includes a lot of redundancy in the spatial domain, the CTU structure may be different from that of normal HEVC contents. Fast CTU partition algorithm is suggested based on entropy of the individual CU [39]. The entropy is quite low in the screen content, because most of area is smooth and the pixel values are mostly equal. Several rules are obtained to terminate CTU partition earlier. For example, if the entropy of a 64 × 64 CU is 0, if the entropy of its four 32 × 32 sub-blocks are equal, or if there are two 32 × 32 sub-blocks that have the same entropy as the other two 32 × 32 sub-blocks, then partition may be terminated. Using this method, the encoding time attained 32% reduction on average, BD rate exhibited 0.8% loss, and PSNR obtained a 0.09% loss, compared to the test mode HM-12.1+RExt-5.1. Fast CU partition decision using machine learning with features that describe CU statistics and sub-CU homogeneity is suggested in [40], achieving 36.8% complexity reduction on average with 3.0% BD-rate increase.

The IBC is a prediction method that finds a matched block within a current frame and sends a block vector (BV) information to the decoder. Since the amount of BV data is significant, the block vector prediction (BVP) is used to reduce the BV data. A block vector difference (BVD), which is calculated by the difference between a BV and a BVP, is coded by the 3rd order exponential Golomb coding. The IBC using these techniques provides a significant coding gain up to 69.39% BD-rate [41] at the cost of computational complexity increased by more than 30% [42]. Therefore, fast algorithm for block vector search in the IBC is considered. One way is to terminate the block vector search early as possible using a threshold when computing SAD between the block and predicted block. The threshold value is defined based on statistical experimental results. The average time savings was 29.23% with BD-rate 0.41% compared to SCM-2.0 [42]. Other algorithms [43] using thresholds and block activities were reported to reduce the block matching time.

HEVC SCC has also adopted the transform skip mode (TSM). Since screen content usually has more regular and sharper edges, prediction method might work well with no transform that may be inefficient and even worsen the coding performance. For these TUs, DCT or DST is skipped and transform_skip_flag (TSF) is signaled to the decoder. Thus, it is based on the similar principle of IBC that sample values are predicted from other samples in the same picture. These two techniques are highly related in terms of statistical occurrence. For example, up to 94.5% of TSM occurs in IBC-coded CUs [44]. Note that coded_block_flag (CBF) is sent for every TU, indicating all transform coefficients are zero, if CBF is set to 0 and any of transform coefficients are non-zero, if it is set to 1. Thus, we can save bits for two flags by careful modification of signaling, such that, for a 4 × 4 TB of IBC-coded CU, TSF is not signaled and CBF plays the same role as TSF. That is, when CBF is 1, it indicates TSM is selected in encoder. By the method and clever adjustment, [44] obtained reduction of 4 × 4 transform block encoding time by 28.2% with a slight increase of BD-rate by 0.22%.

The most probable mode (MPM) is used to reduce bits in the 4×4 intra-prediction instead of nine prediction modes. The encoder estimates the MPM for the current block based on the availability of neighboring blocks. If the MPM is the same as the prediction mode, we need to send only one bit instead of four bits. In screen content, if all boundary samples are exactly having the same value, all samples can be filled by the boundary samples with the MPM index and any rate-distortion optimization (RDO) can be skipped, named simple intra prediction (SIP) in [45]. Direct prediction from boundary samples is possible by introducing single color mode [46] and independent uniform prediction mode [47, 48].

8.5 Visual Quality Assessment

The main objective of image/video coding is to compress data for storage and transmission, while retaining visual quality reasonable to human eye. The simplest way to assess visual quality is peak signal-to-noise ratio (PSNR) that is computed by difference between the original data and reconstructed data. If it is infinite, that means there is no difference and the quality loss does not exist. However, the lossy coding usually results in degradation due to quantization after prediction and transform. Therefore, it is necessary to take out the irrelevant data to human sensitivity that should be defined by certain dedicated models.

8.5.1 Screen Image Quality Assessment

Numerous researches have been performed to develop perceptual quality assessment for images (IQA) that can be classified into three categories depending on the type of contents: natural image quality assessment (NIQA), document image quality assessment (DIQA), and screen image quality assessment (SIQA). It can be further classified into two categories depending on the measurement method: objective and subjective quality assessment. Objective assessment is preferred with its advantage: firstly, they are usually low complexity and secondly, we can classify distortions into several known components such as blocking, ringing, and blurring. Widely used objective assessment metrics are PSNR, SSIM (Structural Similarity) [49], gradient-based [50], image feature-based, and machine learning-based algorithms. There, of course, have been some limitations that they may not be exactly suitable for human observers. Subjective quality assessment is a human judgment-based method. Several test procedures have been defined in

ITU-R Rec. BT.500-11: namely, SS (Single Stimulus), SC (Stimulus Comparison), SSCQE (Single Stimulus Continuous Quality Evaluation), DSIS (Double Stimulus Impairment Scale), SDSCE (Simultaneous Double Stimulus Continuous Evaluation), and DSCQS (Double Stimulus Continuous Quality Scale) [SE2].Another classification is possible depending on the existence of reference images: full-reference, reduced-reference, and no-reference IQA algorithms [51]. The result of quality assessment is reported as either a scalar value or a spatial map denoting the local quality of each image region. Some of best-performing algorithms have been shown to generate quality estimates that correlate with human ratings, typically yielding Spearman rank-order and Pearson linear correlation coefficients in excess of 0.9. These IQA algorithms are summarized in Figure 8.13 [51].

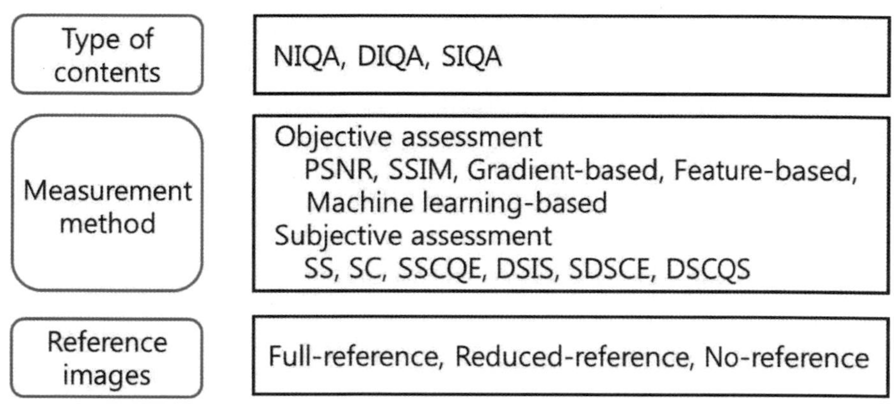

Figure 8.13 Classification of IQA depending on type of images, measurement method, and reference images.

Firstly, NIQA has been studied tremendously during the last several decades. Natural images are usually obtained by visual camera that produces pictorial data. Recently, DIQA has attracted attention in the research community due to the necessity of digitization of old documents or imaged documents that their original features should be maintained. Most DIQA algorithms are designed in no-reference manner, since the original documents may not exist. The effectiveness of DIQA methods can be expressed by accuracy of character recognition. Since SCIs include pictorial regions beside textual regions without environmental degradations, features quite differ from those of the document images and DIQA methods cannot be directly adopted to evaluate the visual quality. The NIQA methods cannot be applied to evaluate the quality of SCIs either. Thus, new screen image database and quality assess metrics have to be developed. In [52], 20 reference and 980 distorted SCIs are included in database that can be downloaded in [53]. Distorted images are generated by applying the typical seven distortions: Gaussian noise, Gaussian blur, motion blur, contrast change, JPEG, JPEG2000 (see Chapter 5), and layer segmentation based coding (LSC) [54] that firstly separates SCIs into textual and pictorial blocks with a segmentation method and applies different encoding method.

8.5.2 Objective Quality Assessment

It has been observed that natural images and textual images have different properties in terms of energy in the spatial frequency domain. To examine this, we decompose images using Fourier transform and then compute energy of the frequency coefficients. Energy of natural images linearly falls off from low to high frequency, while that of textual images has a peak at high frequency, since there are lot of small characters and sharp edges. Thus, SCIs consisting of two or more different contents need to be evaluated by relevant IQA metrics for each content. Since the final decision for quality assessment is to be made for the compound whole images rather than regional images, we still have to develop how to aggregate them.

There are various ways to classify textual and pictorial content, such as gradient-based [55], text detection [56], and segmentation-based, etc. In [57], a block classification approach is suggested by making use of the information content map computed based on the local variance in the 4×4 block. Since textual regions contain high contrast edges, the local information is higher than in pictorial regions. By applying an empirical threshold on the mean of the block information, the textual and pictorial regions can be separated. The quality of each content can be assessed by any methods, although the most popular one would be the SSIM that combines local luminance, contrast and structural similarities. The three types of similarity between the reference and distorted images are pooled into an aggregated quality index. In SCIs, however, some incorrect quality scores happen from the average pooling. Therefore, [58] suggests a structure induced quality metric (SIQM) based on structural degradation model (SDM) defined by

$$\mathrm{SIQM}(\mathbf{r}, \mathbf{d}) = \frac{\sum_{i=1}^{M} \mathrm{SSIM\_MAP}(r_i, d_i) \cdot \mathrm{SDM}(r_i)}{\sum_{i=1}^{M} \mathrm{SDM}(r_i)} \tag{8.14}$$

where **r** and **d** denotes reference and distorted image signal and $\mathrm{SDM}(r_i)$ is defined by

$$\mathrm{SDM}(\mathbf{r}) = 1 - \mathrm{SSIM}(\mathbf{r}, \mathbf{r}_f) \quad (8.15)$$

where $\mathbf{r}_f$ is generated by applying a simple circular-symmetric Gaussian low-pass filter. Distortion maps generated by this method show more highlighted around the texts than in the pictorial regions. Performance of the SIQM is 0.852 on average Spearman and Pearson correlation coefficient, while the SSIM produces 0.750.

Another pooling method is suggested in [57], based on weighted average of textual quality Q_T and pictorial quality Q_P, defined by

$$Q_S = \frac{Q_T \cdot E(\omega_T) + Q_P \cdot E(\omega_P)}{E(\omega_T) + E(\omega_P)} \quad (8.16)$$

where $E(\omega_T)$ and $E(\omega_P)$ denote the expectation of the local energy for the textual and pictorial regions, respectively. These quantities take a role of weighting factor for each content. The higher local energy, the more importance in the region. Q_T and Q_P are computed by another weighted SSIM metric. Performance of this method is 0.851 on average Spearman and Pearson correlation coefficient, while the SSIM produces 0.744, which are similar to those in [58].

8.5.3 Subjective Quality Assessment

Subjective testing methodologies can be roughly categorized into two types: the single stimulus and double stimulus. The former asks the viewers to rate the quality of one distorted image, while the later asks the viewers to rate the quality between reference and distorted images. After testing, mean opinion score (MOS) of ten levels is computed. The higher MOS value is, the more correlation with human eye is. It reveals that the subjective quality scores for SCC is better than that for HEVC [59]. That is, SCC provides better performance than HEVC at the same distortion level as shown in Figure 8.14. However, there are many factors affecting human vision when viewing SCIs, including area ratio and region distribution of textual regions, size of characters, and content of pictorial regions, etc. [60]. When testing by subjects, the consistency of all judgments for each image should be examined. It can be measured by the confidence interval derived from the value and standard deviation of scores. Generally, with a 95% confidence level, the testing scores is regarded as confident.

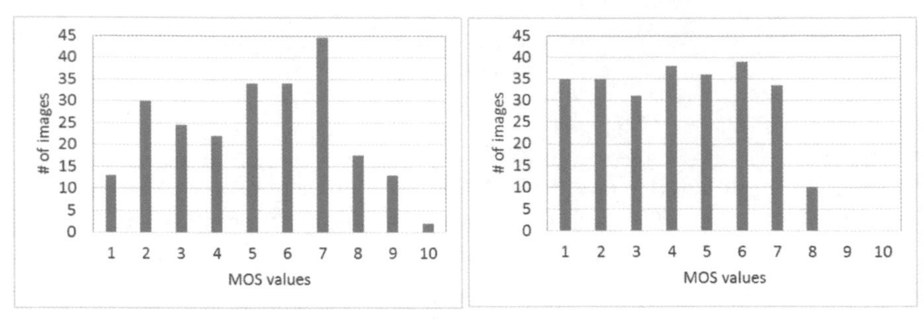

Figure 8.14 Histogram of the MOS values for (left) SCC and (right) HEVC [59]. Higher MOS values are achieved by SCC © IEEE 2015.

8.6 Other SCC Algorithms

8.6.1 Segmentation

Since the SCIs are mixed with text, graphics, and natural pictures, researchers are interested in segmentation of these into several regions and applied different compression algorithms to different image types. In [61], two-step segmentation was developed: block classification and refinement. The first step is to classify 16×16 non overlapping blocks into text/graphics blocks and picture blocks by counting the number of colors in each block. If the number of colors is larger than a certain threshold, the block is classified as picture block. The underlying reason is that natural pictures generally have a large number of colors, while text has a limited number of colors. If the number of colors is more than a threshold, it will be classified into pictorial block, otherwise as text/graphics. In this step, it produces a coarse segmentation, because it may contain different image types. Therefore, a refinement segmentation is followed to extract textual pixels from pictorial pixels. Shape primitives such as horizontal/vertical line or rectangle with the same color are extracted and compared the size and color with some threshold. Thus, it is called shape primitive extraction and coding (SPEC) [61].

In [62], foreground and background separation algorithm is developed. They use the smoothness property of the background and the deviation property of the foreground. The overall segmentation algorithm is summarized as: firstly, if all pixels in the block have the same color, it can be background or foreground by considering neighboring blocks. Second, if all pixels can be predicted with small enough error using least square fitting [63] method, it can be background. Third, they run the segmentation algorithm after decomposing the block size into four smaller blocks until the size 8×8. This algorithm outperforms SPEC in terms of precision and recall.

8.6.2 Rate Control

The rate control is always an important factor to define codec's performance even in SCC. It is required to decide importance of different types of images. The more bit rate for textual regions, the finer quality we obtain for them, while obtaining worse quality for pictorial regions. In video coding it is related to the frame rate. It helps to utilize the bandwidth more efficiently. Rate control can be performed in two procedures: bit allocation and bit control. Once available bits are allocated in GOP level, picture level, and CU level, the next step is to adjust the coding parameters so that the actual amount of bits consumed is close to the pre-allocated target bits. It is desired for a video codec to minimize the bit rate as well as to minimize the distortion which is caused by data compression coding. Thus, the rate control problem is formulated to minimize the distortion D, subject to a rate constraint R to derive the optimal coding parameter P_{opt}:

$$P_{opt} = \arg\min_{P} D \quad s.t. \;\; R \leq B \tag{8.17}$$

where B is the given bit budget. The coding parameter is a set including coding mode, motion estimation, and quantization parameter (QP). Such constrained problem which is too complicated to be solved in real video codec is converted into unconstrained optimization problem, called rate-distortion optimization (RDO) by using Lagrange multiplier λ as

$$P_{opt} = \arg\min_{P} (D + \lambda R) \,. \tag{8.18}$$

λ serves as a weighting factor for the rate constraint and also indicates the slope of the R-D curve [64]. In the practical applications, however, 8.18 is formulated in a simple form such as the quadratic R-D function [65], which has been adopted by most of video coding standards, defined by

$$R = aQ^{-1} + bQ^{-2} \tag{8.19}$$

where Q is quantization scale instead of distortion due to its simplicity.

The work in [64] proposes λ domain rate control for HEVC. The benefits of $R - \lambda$ rate control are: it is more equivalent to finding the distortion on the R-D curve and it can be more precise than adjusting integer QP since λ can take any continuous positive values. It outperforms the R-Q model in 8.19 by 0.55 dB on average.

However, since the screen content has different characteristics, e.g., abrupt changes, a more appropriate rate control scheme is required. The work in [66] proposes an enhanced algorithm based on the $R - \lambda$ model. First, they analyze the complexity of each picture using a sliding window to handle the discontinuities. Then, bits are allocated and the parameters of the model are adjusted. As a result, it decreases the distortion by 2.25% on average and improves the coding efficiency by 5.6%. Another important aspect in screen content is that there are many repeating patterns among pictures and in the same picture as stated earlier. This feature is utilized to introduce the IBC in the screen content coder. Problem is that bit rate at picture level should be maintained at all coding process. The work in [67] proposes weighted rate distortion optimization (WRDO) solution for screen content coding. A weighting factor ω is now applied to 8.18 as

$$P_{opt} = \arg\min_{P} \left(\omega D + \lambda R\right). \tag{8.20}$$

This algorithm was already implemented in HEVC test model, HM-16.2 with uniform distortion weight for all blocks in the picture. ω is kept as 1 in the normal picture which has less influence, while $\omega > 1$ in the more important picture. In case of hierarchical coding structure, pictures in the highest temporal level are never used as reference pictures and $\omega = 1$. As ω is only determined by the coding structure, it is not related to the content. To solve the problem, the block distortion weight should be determined by each block's influence instead of the fixed uniform weight. In [67], block distortion weight is calculated in two ways: inter weight among pictures considering temporal correlations and intra weight within one picture considering the correlations in IBC process. Then, the overall distortion weight is obtained be taking both aspects together. It results in 14.5% of coding gain for the IBBB coding structure at the cost of 2.9% of coding complexity increase.

8.7 Summary

In the context of screen content coding (SCC), various aspects have to be taken into account, due to different characteristics in the natural camera captured content. The SCC is an extension of HEVC standard with several new tools, including IntraBC [72], palette coding, adaptive color transform, and adaptive motion vector resolution, which were discussed in Section 8.2. IntraBC is a kind of motion estimation/compensation used in the natural video coding but it is performed in intra picture, since there are many redundancies in the spatial domain. By using palette mode, we can increase coding gain by sending an index instead of real color value. For screen content, there may exist blocks containing different features of colors, which lead to less correlation. For these blocks, direct coding in the RGB space may be more effective. This gives motivation for adopting adaptive color transform. Motion vector resolution should also be adaptively decided in multi-featured screen content. In some cases, it can be in fractional resolution, while in other cases it should be in integer resolution.

Since there is strong correlation in the spatial domain of screen content, intra prediction is a key factor to be developed. Though it has been developed and standardized in modern video coding standards including H.264 and HEVC, further developments are required for screen contentvideo. For example, sample-based angular intra-prediction with edge prediction is useful for the SCC discussed in Section 8.3.

Development of fast coding algorithms is necessary, since the SCC often requires high computational complexity. Intra coding and motion compensation are two main parts that need to be speeded up as discussed in Section 8.4.

The main objective of image/video coding is to compress data for storage and transmission, while retaining visual quality reasonable to human eye. Thus, screen image quality assessment has to be introduced in the research arena. We discussed and compared quality assessment methods depending on the type of contents: natural images, document images, and screen contentimages in Section 8.5.

Due to different nature in the screen content, many other algorithms are still being developed such as segmentation and rate control as discussed in Section 8.6. A lot of research results are still being introduced in the literature. Some of them appeared in Section 8.8, giving readers updated projects.

8.8 Projects

P.8.1 Tao, et al. [68] proposed a re-sampling technique for template matching that can increase prediction performance at the cost of overhead information: position, index, and value. Among these three, position will consume more number of bits than the other two. As a solution, they applied the similarity of non-zero prediction error of pixel positions. Implement the encoder and find the compression performance.

P.8.2 Zhang, et al. [69] proposed a symmetric intra block copy (SIBC) algorithm in which utilizes symmetric redundancy. They conducted a simple flipping operation either vertically or horizontally on the reference block before it is used to predict the current block. The filpping operation is easy to implement with low cost. They achieved up to 2.3% BD-rate reduction in lossy coding on some sequences that have a lot of symmetric patterns. Please investigate to find the amount of symmetricity in all test sequences that can be downloaded from [15]. Compare the performance to normal IBC using the reference software [16].

P.8.3 Tsang Chan and Siu [70] have developed a fast local search method that can be used for hash based intra block copy mode. Due to the high computational complexity for the IBC [19], they proposed fast local search by checking the hash values of both current block and block candidates. The encoding time is reduced by up to 25% with only negligible bitrate increase. Implement this algorithm and evaluate the performance using the latest reference software.

P.8.4 The Intra Block Copy (IntraBC) tool efficiently encodes repeating patterns in the screen contents as discussed in Section 8.2.1. It is also applicable for coding of natural content video, achieving about 1.0% bit-rate reduction on average. Chen, et al. [71] have worked for further improvements on IntraBC with a template matching block vector and a fractional search method. The gain on natural content video coding increased up to 2.0%, which is not so big, of course, comparing to the efficiency on screen content video. For example, Pang, et al. [72] achieves 43.2% BD rate savings. However, it reveals that there exists some sort of redundancy in the spatial domain of natural video. Carefully design and implement intra prediction with block copy method for natural video content. Evaluate the coding performance in terms of the BD rate.

P.8.5 Fan, et al. [73] have developed quantization parameter offset scheme based on inter-frame correlation, since the inter-frame correlation among adjacent frames in screen content videos is very high. Firstly, they define a measurement of inter-frame correlation and then, quantization parameter offset for successive frames is appropriately adjusted. Number of correlated sub-blocks is counted based on SAD and thresholding between two frames. The more correlation, the larger quantization parameter may be required. The maximum BD rate gain was over 3.8% and the average performance gain is over

2% compared with the reference software. Implement this kind of optimization problem using different correlation measures and block sizes.

P.8.6 Zhao, et al. [74] proposed Pseudo 2D String Matching (P2SM) for screen content coding. Redundancy of both local and non-local repeated patterns is exploited. Different sizes and shapes of the patterns are also considered. They achieved up to 37.7% Y BD rate reduction for a screen snapshot of a spreadsheet. Implement the P2SM and confirm their results.

P.8.7 Sample-based weighted prediction was discussed in Section 8.3.2. Sanchez [75] also proposed sample-based edge prediction based on gradients for lossless screen content coding in HEVC, since a high number of sharp edges causes inefficient coding performance for screen content video using current video coding tools. It is a DPCM-based intra-prediction method that combines angular prediction with gradient-based edge predictor and a DPCM-based DC predictor. Average bit-rate reduction was 15.44% over current HEVC intra-prediction. Implement the edge predictor and apply it to the latest reference software. Describe the advantage obtained by the edge predictor.

P.8.8 The intra coding in the HEVC main profile incorporates several filters for reference samples, including a bi-linear interpolation filter and a smoothing filter [31]. Kang [76] developed adaptive turn on/off filters for inra-prediction method. The decision is based on two criteria: statistical properties of reference samples and information in the compressed domain. For the former, they used Mahalanobis distance for measuring distances between samples and their estimated distributions. For the latter, they used R-D optimization model in the compressed domain. Implement this method using different distance measures and different transforms. Analyze benefits from turning filters on and off. Derive the most efficient transform for coding screen content by taking into account structural information [77].

P.8.9 Natural video has been commonly coded in 4:2:0 sampling format, since the human visual system is less sensitive to chroma. Screen content video, however, is coded in full chroma format, since the downsampling of chroma introduces blur and color shifting. This is because of the anisotropic features in compound contents [78]. Nevertheless, downsampling is required to increase compression ratio as much as possible. S. Wang, et al. [79] proposed an adaptive downsampling for chroma based on local variance and luma guided chroma filtering [80]. Coding performance is measured in terms of PSNR and SSIM. Implement their adaptive dawonsampling scheme using screen content reference software. Compare the performance to that using full chroma format in terms of different performance criteria including BD rate.

P.8.10 Three transform skip modes (TSM) have been proposed during the development of HEVC, including vertical transform skipping, horizontal transform skipping, and all 2D transform skipping. Vertical transform skipping means applying only the vertical 1D transform and skipping the horizontal 1D transform. This is called 1D TSM. If the prediction errors have minimal correlation in one or both directions, transform doesn't work well. Thus, it can be skipped. The 1D TSM is efficient in the screen content which has strong correlations in one direction but not the other. J.-Y. Kao, et al. [81] developed the 1D TSM based on dynamic range control, modification of the scan order, and coefficient flipping. D. Flynn, et al. [82] discusses more on TSM. Evaluate their effectiveness using the latest SCM software.

P.8.11 Implement and compare the three intra coding techniques: intra string copy [83], intra block copy [84, 20], and intra line copy [85].

P.8.12 Just Noticeable Difference (JND) has been used for measuring image/video distortions. Wang, et al. [86] developed the JND modeling to be used for compressing screen content images. Each edge profile is decomposed into luminance, contrast, and structure, and then evaluate the visibility threshold in different ways. The edge luminance adaptation, contrast masking, and structural distortion sensitivity are also studied. Develop the JND model for lossless and lossy coding of screen content video and evaluate it in terms of human visual sensitivity.

P.8.13 Some fast algorithms for coding screen content are discussed in Section 8.4, mainly dealing with intra coding and motion compensation. Another approach is proposed by Zhang, et al. [87] to speed up intra mode decision and block matching for IntraBC. If we could take proper background detection, encoding time can be saved by skipping mode decision process in the region. Background region can be detected by some sort of segmentation technique discussed in Section 8.6.1. Derive the background detection algorithms by your own means and apply it for coding screen content.

P.8.14 F. Duanmu, et al. [88] developed a transcoding framework to efficiently bridge the HEVC standard (Chapter 5) and it's SCC extension. It can achieve an average of 48% re-encoding complexity reduction with less than 2.14% BD-rate increase. It is designed as a pre-analysis module before intra frame mode selection. Coding modes are exchanged based on statistical study and machine learning. Implement this type of transcoding tool and confirm their results.

P.8.15 C. Chen, et al. [89] proposed a staircase transform coding scheme for screen content video coding, that can be integrated into a hybrid coding scheme in conjunction with conventional DCT. Candidates for the staircase transform include Walsh-Hadamard transform and Haar transform [90]. The proposed approach provides an average of 2.9% compression performance gains in terms of BD-rate reduction. Implement this hybrid coder and evaluate the performance.

References

[1] D. Flynn, J. Sole and T. Suzuki, High efficiency video coding (HEVC) range extensions text specification, Draft 4, JCT-VC, Retrieved 2013-08-07.

[2] I. E. G. Richardson, H.264 and MPEG-4 video compression: Video coding for Next Generation Multimedia, New York: Wiley, 2003.

[3] S. Kodpadi, "Evaluation of coding tools for screen content in High Efficiency Video Coding," Dec. 2015. http://www.uta.edu/faculty/krrao/dip. Click on courses and then click on EE5359. Scroll down and go to Thesis/Project Title and click on S. Kodpadi.

[4] N. Mundgemane , "Multi-stage prediction scheme for screen content based on HEVC," *Master's thesis at University of Texas at Alrington,* 2015. http://www.uta.edu/faculty/krrao/dip. Click on courses and then click on EE5359. Scroll down and go to Thesis/Project Title and click on N. Mundgemane.

[5] H. Yu, et al., "Requirements for an extension of HEVC for coding on screen content," *ISO/IEC JTC 1/SC 29/WG 11 Requirements subgroup,* Jan. 2014.

[6] J. Xu, R. Joshi, and R. A. Cohen, "Overview of the Emerging HEVC Screen Content Coding Extension," *IEEE Trans. CSVT,* vol. 26, no. 1, pp. 50–62, Jan. 2016.

[7] C. Lan, et al., "Intra and inter coding tools for screen contents," *JCTVC-E145,* Mar. 2011.

[8] D. Milovanovic and Z. Bojkovic, "Development of mixed screen content coding technology: SCC-HEVC Extensions Framework," *Proc. of the 13th Int. Conf. on Data Networks, Communications, Computers (DNCOCO'15),* pp. 44–50, Budapest, Hungary, Dec. 2015.

[9] ISO/IEC JTC1/SC29/WG11 and ITU-T SG16 Q6/16, "Joint Call for Proposals for Coding of Screen Content," 17 Jan. 2014.

[10] H. Yu, et al., "Common conditions for screen content coding tests," *JCTVC – U1015-r2,* June 2015.

[11] ISO/IEC JTC1/SC29/WG11 N14399, "Results of CfP on Screen Content Coding Tools for HEVC," Apr. 2014.

[12] ISO/IEC JTC1/SC29/WG11 N15779, "HEVC Screen Content Coding Test Model 6 (SCM 6)," Oct. 2015.

[13] R. Joshi, et al., "High Efficiency Video Coding (HEVC) Screen Content Coding: Draft 5," *JCTVC-V1005,* Oct. 2015.

[14] Access to JCT-VC documents, "http://phenix.int-evry.fr/jct/," [Online].

[15] Test sequences for SCC, "http://pan.baidu.com/share/link?shareid=3128894651&uk=889443731," [Online].

[16] HEVC SCC Extension Reference Software, "https://hevc.hhi.fraunhofer.de/svn/svn_HEVCSoftware/tags/HM-16.8+SCM-7.0/," [Online].

[17] HEVC SCC Software reference manual, "https://hevc.hhi.fraunhofer.de/svn/svn_HEVCSoftware/branches/HM-SCC-extensions/doc/software-manual.pdf," [Online].

[18] Y.-J. Ahn, et al., "Analysis of Screen Content Coding Based on HEVC," *IEIE Trans. on Smart Processing and Computing,* vol. 4, no. 4, pp. 231–236, Aug. 2015.

[19] W. Zhu, et al., "Hash-Based Block Matching for Screen Content Coding," *IEEE Trans. on Multimedia,* vol. 17, no. 7, pp. 935–944, July 2015.

[20] X. Xu, et al., "Block Vector Prediction for Intra Block Copying in HEVC Screen Content Coding," *IEEE Data Compression Conference (DCC),* pp. 273–282, 2015.

[21] L. Guo, et al., "Color palette for screen content coding," *IEEE ICIP,* pp. 5556–5560, Oct. 2014.

[22] X. Xiu, et al., "Palette-based Coding in the Screen Content Coding Extension of the HEVC Standard," *IEEE Data Compression Conference (DCC 2015),* pp. 253–262, April 2015.

[23] L. Zhang, et al., "Adaptive Color-Space Transform for HEVC Screen Content Coding," *IEEE Data Compression Conference (DCC),* pp. 233–242, 2015.

[24] H. S. Malvar, G. J. Sullivan, and S. Srinivasan, "Lifting-based reversible color transformations for image compression," *SPIE Applications of Digital Image Processing,* vol. 7073, Aug. 2008.

[25] M. Wien, High efficiency video coding: Coding tools and specification, Springer, 2015.

[26] Y.-L. Lee, K.-H. Han, and G. J. Sullivan, "Improved lossless intra coding for H.264/MPEG-4 AVC," *IEEE Trans. on Image Processing,* vol. 15, no. 9, pp. 2610–2615, Sept. 2006.

[27] S. Lee, I.-K. Kim and C. Kim, "Residual DPCM for HEVC lossless coding," *JCTVC-M0079,* 13th JCT-VC meeting, Incheon, KR, Apr. 2013.

[28] K. Kim, J. Jeong, and G, Jeon, "Improved residual DPCM for HEVC lossless coding," *27th SIBGRAPI Conf. on Graphics, Patterns and Images,* pp. 95–102, Aug. 2014.

[29] M. Naccari, et al., "Improving inter prediction in HEVC with residual DPCM for lossless screen content coding," *IEEE PCS,* pp. 361–364, San Jose, CA, Dec. 2013.

[30] E. Wige, et al., "Sample-based weighted prediction with directional template matching for HEVC lossless coding," *IEEE PCS,* pp. 305–308, Dec. 2013.

[31] V. Sze, M. Budagavi, and G. J. Sullivan, High efficiency video coding (HEVC): Algorithms and architectures, Springer, 2014.

[32] M. Zhou, et al., "HEVC lossless coding and improvements," *IEEE Trans. CSVT,* vol. 22, no. 12, pp. 1839–1843, Dec. 2012.

[33] Y. H. Tan, C. Yeo and Z. Li, "Residual DPCM for lossless coding in HEVC," *Proc. IEEE Int. Conf. Acoustics, Speech Signal Processing (ICASSP),* pp. 2021–2025, May 2013.

[34] V. Sanchez, et al., "HEVC-based Lossless Compression of Whole Slide Pathology Images," *Proc. 2014 IEEE Global Conference on Signal and Information Processing (GlobalSIP),* pp. 452–456, Dec. 2014.

[35] V. Sanchez, "Lossless screen content coding in HEVC based on sample-wise median and edge prediction," *IEEE International Conference on Image Processing (ICIP),* pp. 4604–4608, 2015.

[36] V. Sanchez, "Fast intra-prediction for lossless coding of screen content in HEVC," *2015 IEEE Global Conference on Signal and Information Processing (GlobalSIP),* pp. 1367–1371, 2015.

[37] B. Li and J. Xu, "A Fast Algorithm for Adaptive Motion Compensation Precision in Screen Content Coding," IEEE *Data Compression Conference (DCC),* pp. 243–252, 2015.

[38] B. Li, J. Xu, and F. Wu, "A Unified Framework of Hash-based Matching for Screen Content Coding," *IEEE VCIP'14*, pp. 530–533, Dec. 7–10, 2014.

[39] M. Zhang, Y. Guo, and H. Bai, "Fast intra partition algorithm for HEVC screen content coding," *IEEE Visual Communications and Image Processing Conf.*, pp. 390–393, 2014.

[40] F. Duanmu, Z. Ma, and Y. Wang, "Fast CU partition decision using machine learning for screen content compression," *IEEE Int. Conf. on Image Processing (ICIP)*, pp. 4972–4976, 2015.

[41] G. Bjontegaard, "Calculation of average PSNR differences between RD-curves," *ITU-T SG 16 Q 6, VCEG-M33*, Mar. 2001.

[42] J. Ma and D. Sim, "Early Termination of Block Vector Search for Fast Encoding of HEVC Screen Content Coding," *IEIE Trans. on Smart Processing and Computing*, vol. 3, no. 6, pp. 388–392, Dec. 2014.

[43] D.-K. Kwon and M. Budagavi, "Fast intra block copy (intra BC) search for HEVC Screen Content Coding," *IEEE ISCAS*, pp. 9–12, June 2014.

[44] D. Lee, et al., "Fast transform skip mode decision for HEVC screen content coding," *IEEE Int. Symp. on Broadband Multimedia Systems and Broadcasting (BMSB)*, pp. 1–4, 2015.

[45] S.-H. Tsang, Y.-L. Chan and W.-C. Siu, "Fast and Efficient Intra Coding Techniques for Smooth Regions in Screen Content Coding Based on Boundary Prediction Samples," *IEEE ICASSP*, pp. 1409–1413, Brisbane, Australia, Apr. 2015.

[46] Y.-W. Chen, et al., "SCCE3 Test D.1: Single color intra mode for screen content coding," *JCTVC of ITU-T SG 16 WP 3 and ISO/IEC JTC1/SC29/WG11, JCTVCR0058, Sapporo, Japan*, June–July 2014.

[47] X. Zhang, R. A. Cohen and A. Vetro, "Independent Uniform Prediction mode for screen content video coding," *IEEE Visual Communications and Image Processing Conference*, pp. 129–132, 7–10 Dec. 2014.

[48] X. Zhang, and R. Cohen, "SCCE3 Test D.2: Independent Uniform Prediction (IUP) Mode," *JCT-VC of ITU-T SG 16 WP 3 and ISO/IEC JTC1/SC29/WG11, JCTVC-R0200, Sapporo, Japan*, June–July 2014.

[49] Z. Wang, et al., "Image quality assessment: From error visibility to structural similarity," *IEEE Trans. on Image Processing*, vol. 13, no. 4, pp. 600–612, Apr. 2004.

[50] J. Zhu and N. Wang, "Image quality assessment by visual gradient similarity," in *Proceedings of the IEEE Trans. on Image Processing*, vol. 21, no. 3, pp. 919–933, Mar. 2012.

[51] D. M. Chandler, "Seven challenges in image quality assessment: Past, present, and future research," *ISRN Signal Process.,* vol. 2013, pp. 1–53, 2013.

[52] H. Yang, Y. Fang and W. Lin, "Perceptual Quality Assessment of Screen Content Images," *IEEE Trans. on Image processing,* vol. 24, no. 11, pp. 4408–4421, Nov. 2015.

[53] SIQAD (Perceptual Quality Assessment of Screen Content Images), "https://sites.google.com/site/subjectiveqa/," [Online].

[54] Z. Pan, et al., "A low-complexity screen compression scheme for interactive screen sharing," *IEEE Trans, CSVT,* vol. 23, no. 6, pp. 949–960, June 2013.

[55] S. Wang, et al., "Content-aware layered compound video compression," *IEEE ISCAS,* pp. 145–148, 2012.

[56] A. Risnumawan, et al., "A robust arbitrary text detection system for natural scene images," *Expert Systems with Applications,* vol. 41, no. 18, pp. 8027–8048, Dec. 2014.

[57] S. Wang, et al., "Perceptual screen content image quality assessment and compression," *IEEE Int. Conf. on Image Processing (ICIP),* pp. 1434–1438, 2015.

[58] K. Gu, et al., "Screen image quality assessment incorporating structural degradation measurement," *IEEE Int. Symp. on Circuits and Systems (ISCAS),* pp. 125–128, 2015.

[59] S. Shi, et al., "Study on subjective quality assessment of Screen Content Images," *IEEE Picture Coding Symposium (PCS),* pp. 75–79, 2015.

[60] H. Yang, et al., "Subjective quality assessment of screen content content images," *International Workshop on Quality of Multimedia Experience (QoMEX),* pp. 257–262, 2014.

[61] T. Lin and P. Hao, "Compound Image Compression for Real-Time Computer Screen Image Transmission," *IEEE Trans. on Image Processing,* vol. 14, no. 8, pp. 993–1005, Aug. 2005.

[62] S. Minaee and Y. Wang, "Screen content image segmentation using least absolute deviation fitting," *IEEE Int. Conf. on Image Processing (ICIP),* pp. 3295–3299, 2015.

[63] Y. Li and G. R. Arce, "A maximum likelihood approach to least absolute deviation regression," *EURASIP Journal on Applied Signal Processing,* pp. 1762–1769, Sept. 2004.

[64] B. Li, et al., "Lambda Domain Rate Control Algorithm for High Efficiency Video Coding," *IEEE Trans. on Image Processing,* vol. 23, no. 9, pp. 3841–3854, Sept. 2014.

[65] T. Chiang and Y.-Q. Zhang, "A new rate control scheme using quadratic rate distortion model," *IEEE Trans. on CSVT,* vol. 7, no. 1, pp. 246–250, Jan. 1997.

[66] Y. Guo, et al., "Rate control for screen content coding in HEVC," *IEEE Int. Symp. on Circuits and Systems (ISCAS),* pp. 1118–1121, 2015.

[67] W. Xiao, et al., "Weighted rate-distortion optimization for screen content coding," *IEEE Trans. on CSVT,* 2016 (Early Access).

[68] P. Tao, et al., "Improvement of re-sample template matching for lossless screen content video," *IEEE International Conference on Multimedia and Expo (ICME),* pp. 1–6, 2015.

[69] K. Zhang, et al., "Symmetric intra block copy in video coding," *IEEE International Symposium on Circuits and Systems (ISCAS),* pp. 521–524, 2015.

[70] S.-H. Tsang, Y.-L. Chan, and W.-C. Siu, "Hash based fast local search for Intra Block Copy (IntraBC) mode in HEVC screen content coding," *Asia-Pacific Signal and Information Processing Association Annual Summit and Conference (APSIPA),* pp. 396–400, 2015.

[71] H. Chen, et al., "Improvements on Intra Block Copy in natural content video coding," *IEEE International Symposium on Circuits and Systems (ISCAS),* pp. 2772–2775, 2015.

[72] C. Pang, et al., "Intra Block Copy for HEVC Screen Content Coding," *IEEE Data Compression Conference (DCC),* p. 465, 2015.

[73] J. Fan, et al., "Inter-frame Correlation Based Quantization Parameter Offset Optimization for Screen Content Video Coding," *IEEE International Conference on Multimedia Big Data (BigMM),* pp. 420–423, 2015.

[74] L. Zhao, et al., "Pseudo 2D String Matching Technique for High Efficiency Screen Content Coding," *IEEE Transactions on Multimedia,* vol. 18, no. 3, pp. 339–350, 2016.

[75] V. Sanchez, "Sample-based edge prediction based on gradients for lossless screen content coding in HEVC," *IEEE Picture Coding Symposium (PCS),* pp. 134–138, 2015.

[76] J.-W. Kang, "Sample selective filter in HEVC intra-prediction for screen content video coding," *The Institute of Engineering and Technology (IET) Journals & Magazines,* vol. 51, no. 3, pp. 236–237, Feb. 2015.

[77] J.-W Kang, "Structured sparse representation of residue in screen content video coding," *Electronics Letters,* vol. 51, no. 23, pp. 1871–1873, 2015.

[78] C. Lan, G. Shi, and F. Wu, "Compress compound images in H.264/MPGE-4 AVC by exploiting spatial correlation," *IEEE Trans. on Image Processing,* vol. 19, no. 4, pp. 946–957, 2010.
[79] S. Wang, et al., "Joint Chroma Downsampling and Upsampling for screen Content Image," *IEEE Trans. on CSVT,* vol. 26, no. 9, pp. 1595–1609, Sept. 2016.
[80] T. Vermeir, et al., "Guided Chroma Reconstruction for Screen Content Coding," *IEEE Trans. on CSVT,* vol. 26, no. 10, pp. 1884–1892, Oct. 2016.
[81] J.-Y. Kao, et al., "Improved transform skip mode for HEVC screen content coding," *Int. Conf. on Image Processing Theory, Tools and Applications (IPTA),* pp. 504–509, 2015.
[82] D. Flynn, et al., "Overview of the Range Extensions for the HEVC Standard: Tools, Profiles, and Performance," *IEEE Trans. on CSVT,* vol. 26, no. 1, pp. 4–19, Jan 2016.
[83] F. Zou, et al., "Hash Based Intra String Copy for HEVC Based Screen Content Coding," *IEEE ICME,* Torino, Italy, June 2015.
[84] X. Xu, et al., "Intra Block Copy in HEVC Screen Content Coding Extensions," *IEEE J. of Emerging Selected Topics in Circuits and Systems (Early Access).*
[85] C. C. Chen and W. H. Peng, "Intra Line Copy for HEVC Screen Content Coding," *IEEE Trans. on CSVT,* vol. 27, pp. 1568–1579, July 2017.
[86] S. Wang, et al., "Just Noticeable Difference Estimation for Screen Content Images," *IEEE Trans. on Image Processing,* vol. 25, no. 8, pp. 3838–3851, 2016.
[87] H. Zhang, et al., "Fast intra mode decision and block matching for HEVC screen content compression," *IEEE ICASSP,* pp. 1377–1381, 2016.
[88] F. Duanmu, et al., "A novel screen content fast transcoding framework based on statistical study and machine learning," *IEEE ICIP,* pp. 4205–4209, 2016.
[89] C. Chen, et al., "A staircase transform coding scheme for screen content video coding," *IEEE ICIP,* pp. 2365–2369, 2016.
[90] K. R. Rao and P. C. Yip, The transform and data compression handbook, CRC Press, 2001.
[91] J. Li, et al., "Diversity-Based Reference Picture Management for Low Delay Screen Content Coding," *IEEE Trans.* CSVT, (Early Access).
[92] W. Xiao, et al., "Fast Hash-based Inter Block Matching for Screen Content Coding," *IEEE Trans.* CSVT, (Early Access).

[93] S. Wang, et al., "Reduced-Reference Quality Assessment of Screen Content Images," *IEEE Trans. CSVT,* (Early Access).

[94] K. Zhang, et al., "Segmental Prediction for Video Coding," *IEEE Trans. CSVT,* (Early Access).

[95] W. S. Kim, et al., "Cross-Component Prediction in HEVC", *IEEE Trans. CSVT,* (Early Access).

The Reader is Referred to Chapter 5 where References Related to SCC are Added.

H.264 Advance Video Coding (AVC)/MPEG-4 Part 10 References

H1 R. Schäfer, T. Wiegand and H. Schwarz, "The emerging H.264/AVC standard," *EBU Technical Review*, pp. 1–12, Jan. 2003.

H2 JVT Draft ITU-T recommendation and final draft international standard of joint video specification (ITU-T Rec. H.264–ISO/IEC 14496-10 AVC), March 2003, JVT-G050 available on http://ip.hhi.de/imagecom_G1/assets/pdfs/JVT-G050.pdf.

H3 T. Wiegand *et al.*, "Overview of the H.264/AVC video coding standard," *IEEE Trans. CSVT*, Special Issue on H.264/AVC, vol. 13, pp. 560–576, July 2003.

H4 H.S. Malvar et al., "Low-complexity transform and quantization in H.264/AVC," IEEE Trans. CSVT, vol. 13, pp. 598–603, July 2003.

H5 T. Stockhammer, M.M. Hannuksela and T. Wiegand, "H.264/AVC in wireless environments," IEEE Trans. CSVT, vol. 13, pp. 657–673, July 2003.

H6 S. Wenger, "H.264/AVC over IP," IEEE Trans. CSVT, vol. 13, pp. 645–656, July 2003.

H7 A. Tamhankar and K.R. Rao, "An overview of H.264/MPEG-4 part 10," Proc. 4th EURASIP-IEEE Conf. focused on Video/Image Processing and Multimedia Communications, vol. 1, pp. 1–51, Zagreb, Croatia, July 2003.

H8 J. Ostermann et al., "Video coding with H.264/AVC: Tools, performance, and complexity", IEEE Circuits and Systems Magazine, vol. 4, pp. 7–28, First Quarter 2004.

H9 M. Fieldler, "Implementation of basic H.264/AVC decoder," Seminar Paper at Chemnitz University of Technology, June 2004.

H10 Y. Zhang et al., "Fast 4x4 intra-prediction mode selection for H.264", in Proc. Int. Conf. Multimedia and Expo, pp. 1151–1154, Taipei, Taiwan, Jun. 2004.

H11 G.J. Sullivan, P. Topiwala and A. Luthra, "The H.264/AVC advanced video coding standard: Overview and introduction to the fidelity range extensions," *SPIE Conf. on Applications of Digital Image Processing XXVII*, vol. 5558, pp. 53–74, Aug. 2004.

H12 F. Fu et al., "Fast intra prediction algorithm in H.264/AVC", Proc. 7th Int. Conf. Signal Process., pp. 1191–1194, Beijing, China, Sep. 2004.

H13 A. Puri, X. Chen and A. Luthra, "Video coding using the H.264/MPEG-4 AVC compression standard," *Signal Processing: Image Communication*, vol. 19, pp. 793–849, Oct. 2004.

H14 F. Pan et al., "Fast intra mode decision algorithm for H.264/AVC video coding," Proc. IEEE ICIP, pp. 781–784, Singapore, Oct. 2004.

H15 Proc. IEEE, Special Issue on Advances in Video Coding and Delivery, vol. 93, pp. 3–193, Jan. 2005. This has several overview papers.

H16 T. Sikora, "Trends and perspectives in image and video coding," *Proc. IEEE*, vol. 93, pp. 6–17, Jan. 2005.

H17 G.J. Sullivan and T. Wiegand, "Video compression – From concepts to H.264/AVC standard", (Tutorial), *Proc. IEEE*, vol. 93, pp. 18–31, Jan. 2005.

H18 D. Marpe and T. Wiegand, "H.264/MPEG4-AVC Fidelity Range Extensions: Tools, Profiles, Performance, and Application Areas", IEEE ICIP, vol. 1, pp. I–596, 11–14 Sept. 2005.

H19 Y.-L. Lee and K.-H. Han, "Complexity of the proposed lossless intra for 4:4:4", (ISO/IEC JTC1/SC29/WG11 and ITU-T SG 16 Q.6) document JVT-Q035, 17–21 Oct. 2005.

H20 D. Kumar, P. Shastry and A. Basu, "Overview of the H.264/AVC," 8th Texas Instruments Developer Conference India, 30 Nov.–1 Dec. 2005, Bangalore, India.

H21 Proc. IEEE, Special Issue on Global Digital Television: Technology and Emerging Services, vol. 94, Jan. 2006.

H22 G. Raja and M. Mirza, "In-loop de-blocking filter for H.264/AVC Video," Proc. the 2nd IEEE-EURASIP Int. Symp. Commun., Control and Signal Processing 2006, ISCCSP, Marrakech, Morocco, March 2006.

H23 S.-K. Kwon, A. Tamhankar and K.R. Rao, "Overview of H.264/MPEG-4 Part 10," *J. Visual Commun. Image Representation (JVCIR),* vol. 17, pp. 186–216, April 2006. Special Issue on "Emerging H.264/AVC Video Coding Standard".

H24 S. Kumar *et al.*, "Error resiliency schemes in H.264/AVC standard," *J. Visual Commun. Image Representation (JVCIR),* Special Issue on H.264/AVC, vol. 17, pp. 425–450, April 2006.

H25 D. Marpe, T. Wiegand and G. J. Sullivan, "The H.264/MPEG-4 AVC standard and its applications," *IEEE Communications Magazine*, vol. 44, pp. 134–143, Aug. 2006.

H26 S. Naito and A. Koike, "Efficient coding scheme for super high definition video based on extending H.264 high profile," in Proc. SPIE Visual Commun. Image Process., vol. 6077, pp. 607727-1-607727-8, Jan. 2006.

H27 J. Kim et al., "Complexity reduction algorithm for intra mode selection in H.264/AVC video coding" J. Blanc-Talon et al. (Eds.): pp. 454–465, ACIVS 2006, LNCS 4179, Springer, 2006.

H28 S. Ma and C.C. J. Kuo, "High-definition video coding with super macroblocks," in Proc. SPIE VCIP, vol. 6508, pp. 650816-1-650816-12, Jan. 2007.

H29 T. Wiegand and G.J. Sullivan, "The H.264/AVC video coding standard," *IEEE SP Magazine*, vol. 24, pp. 148–153, March 2007. (Info on H.264/AVC resources).

H30 H.264/AVC JM 18.0 reference software: http://iphome.hhi.de/suehring/tml/download/

H31 "H.264 video compression standard," White paper, Axis communications.

H32 W. Lee *et al.*, "High speed intra prediction scheme for H.264/AVC," IEEE Trans. Consumer Electronics, vol. 53, no. 4, pp. 1577–1582, Nov. 2007.

H33 Y.Q. Shi and H. Sun, "Image and video compression for multimedia engineering," Boca Raton: CRC Press, II Edition (Chapter on H.264), 2008.

H34 B. Furht and S.A. Ahson, "Handbook of mobile broadcasting, DVB-H, DMB, ISDB-T and MEDIAFLO," Boca Raton, FL: CRC Press, 2008. (H.264 related chapters)

H35 M. Jafari and S. Kasaei, "Fast intra- and inter-prediction mode decision in H.264 advanced video coding," International Journal of Computer Science and Network Security, vol. 8, pp. 130–140, May 2008.

H36 A.M. Tourapis (January 2009), "H.264/14496-10 AVC reference software manual" [Online]. Available: http://iphome.hhi.de/suehring/tml/JM%20Reference%20Software%20Manual%20%28JVT-AE010%29.pdf

H37 P. Carrillo, H. Kalva, and T. Pin, "Low complexity H.264 video encoding," SPIE, vol. 7443, paper # 74430A, Aug. 2009.

H38 B.M.K. Aswathappa and K.R. Rao, "Rate-distortion optimization using structural information in H.264 strictly intra-frame encoder," IEEE Southeastern Symp. on System Theory (SSST), Tyler, TX, March 2010.

H39 D. Han *et al.*, "Low complexity H.264 encoder using machine learning", IEEE SPA 2010, pp. 40–43, Poznan, Poland, Sept. 2010.

H40 T. Sathe, "Complexity reduction of H.264/AVC motion estimation using OpenMP," M.S. Thesis, EE Dept., University of Texas at Arlington, Arlington, Texas, May 2011. http://www.uta.edu/faculty/krrao/dip click on courses and then click on EE5359 Scroll down and go to Thesis/Project Title and click on T. Sathe.

H41 D. Han, A. Kulkarni and K.R. Rao, "Fast inter-prediction mode decision algorithm for H.264 video encoder," IEEE ECTICON 2012, Cha Am, Thailand, May 2012.

H42 H.264/MPEG-4 AVC: http://en.wikipedia.org/wiki/H.264

H43 S. Subbarayappa, "Implementation and analysis of directional discrete cosine transform in H.264 for baseline profile," M.S. Thesis, EE Dept., UTA, May 2012.

H44 I.E. Richardson, "H.264/MPEG-4 Part 10 White Paper," www.vcodex.com

H45 I.E. Richardson, "Overview: What is H.264?" 2011. www.vcodex.com

H46 I.E. Richardson, "A Technical Introduction to H.264/AVC," 2011. [Online]. Available: http://www.vcodex.com/files/H.264_technical_introduction.pdf

H47 I.E. Richardson, "White Paper: H.264/AVC intra prediction," 2011. www.vcodex.com

H48 I.E. Richardson, "White Paper: H.264/AVC inter prediction," 2011. www.vcodex.com

H49 I.E. Richardson, "White Paper: 4×4 transform and quantization in H.264/AVC," Nov. 2010. www.vcodex.com

H50 I.E. Richardson, "White Paper: H.264/AVC loop filter," 2011. www.vcodex.com

H51 Joint Video Team (JVT), ITU-T website, http://www.itu.int/ITU-T/studygroups/com16/jvt/

H52 "RGB Spectrum's H.264 codecs hit the mark in the latest U.S. missile test program," May 2012. www.rgb.com. See [B18].

H53 ISO website: http://www.iso.org/iso/home.htm

H54 IEC website: http://www.iec.ch/

H55 ITU-T website: http://www.itu.int/ITU-T/index.html

H56 C.-L. Su, T.-M. Chen and C.-Y. Huang," Cluster-Based Motion Estimation Algorithm with Low Memory and Bandwidth Requirements for H.264/AVC Scalable Extension", IEEE Trans. CSVT, vol. 24, pp. 1016–1024, June 2014.

H.264/14496-10 Advanced video coding reference software manual 28 June–3 July, 2009 JM 16.0
http://iphome.hhi.de/suehring/tml/JM%20Reference%20Software%20Manual%20(JVT-AE010).pdf
H.264/mpeg-4 AVC reference software joint Model 18.4 available at http://iphome.hhi.de/suehring/tml/download/jml8.4.zip

H57 FFmpeg Developers (2013): Libavcodec H.264 Decoder [Online]. Available: http://ffmpeg.org, accessed Aug 1, 2014.

H58 X264: A free H.264 Encoder [Online] Available: http://www.videolan.org/developers/x.264.html, accessed Aug.1, 2014.

H59 D. S. Eagle, "Beyond the Full HD (UHD & 8K)," Sunday 12 April 2015, 3:30 PM to 4:45 PM, NAB Show, Las Vegas, NV, April 2015.

Sony's XAVC format employs MPEG-4 AVC/H.264 level 5.2, the highest picture resolution and frame rate video compression codec based on industry standards. XAVC enables a very wide range of operational possibilities for content production, notably: from Proxy to 4K pixel resolutions, intra frame and long GOP schemes, and 1080 50P/60P infrastructure capability.
Built with the principles of workflow efficiency, evolution and optimized image quality at its heart, Sony's XAVC can support the following content formats:

- 4K (4096 x 2160 and 3840 x 2160), HD and proxy resolution
- MPEG-4 AVC/H.264 video compression
- 12, 10 and 8 bit color depth
- Up to 60fps
- MXF wrapping format can be used
- 4:4:4, 4:2:2, 4:2:0 color sampling

XAVC has been developed as an open format, providing a license program for other manufacturers in the broadcast and production industry to develop their own high quality and high frame rate products.
Website: http://www.sony.co.uk/pro/article/broadcast-products-sonys-new-xavc-recording-format-accelerates

H60 I. E. G. Richardson, "The H.264 Advanced Video Compression Standard," II Edition, Wiley, 2010.

H61 T. Wiegand et al., "Introduction to the Special Issue on Scalable Video Coding – Standardization and Beyond," IEEE Trans. CSVT, vol. 17, pp. 1099–1269, Sept. 2007.

H62 K. Muller et al.," Multi-view video coding based on H.264/AVC using hierarchical B-frames," in Proc. PCS 2006, Picture Coding Symp., Beijing, China, April 2006.

H63 T. Wiegand et al., "Joint draft 5: scalable video coding," Joint Video Team of ISO/IEC MPEG and ITU-T VCEG, Doc. JVT-R201, Bangkok, Jan. 2006.

H64 G. J. Sullivan, " The H.264/MPEG4-AVC video coding standard and its deployment status," in proc. SPIE conf. visual communications and image processing (VCIP), Beijing, China, July 2005.

H65 T. Wiegand et al., "Rate-constrained coder control and comparison of video coding standards," IEEE Trans. CSVT, vol. 13, pp. 688–703, July 2003.

H66 C. Fenimore et al., " Subjective testing methodology in MPEG video verification," in Proc. SPIE Applications of Digital Image Processing, vol. 5558, pp. 503–511, Denver, CO, Aug. 2004.

H67 MPEG industry forum resources [online] Available: http://www.mpegif.org/resources.php

H68 H. Schwarz, D. Marpe and T. Wiegand, "Overview of the scalable H.264/AVC extension," in Proc. IEEE Int. Conf. Image Processing, Atlanta, GA, pp. 161–164, Oct. 2006.

H69 C.-H. Kuo, L.-L. Shih and S.-C. Shih, "Rate control via adjustment of Lagrange multiplier for video coding", IEEE Trans. CSVT, (EARLY ACCESS)

H70 A.K. Au-Yeung, S. Zhu and B. Zeng, 'Partial video encryption using multiple 8x8 transforms in H.264 and MPEG-4,' Proceedings of ICASSP 2011, pp. 2436–2439, Prague, Czech Republic., May 2011.

H71 A.K. Au-Yeung, S. Zhu and B. Zeng, 'Design of new unitary transforms for perceptual video encryption,' IEEE Trans. CSVT, vol. 21, pp. 1341–1345, Sept. 2011.

H72 B. Zeng et al., 'Perceptual encryption of H.264 videos: embedding sign-flips into the integer-based transforms,' IEEE Trans. Information Forensics and Security, vol. 9, no. 2, pp. 309–320, Feb. 2014.

H73 S.J. Li et al., "On the design of perceptual MPEG-video encryption algorithms," IEEE Trans. CSVT, vol. 17, pp. 214–223, Feb. 2007.

H74 B.Y. Lei, K.T. Lo and J. Feng, 'Encryption Techniques for H.264 Video," The Handbook of MPEG Applications: Standards in Practice, pp. 151–174, Edited by M.C. Angelides and H. Agius, Wiley, Feb. 2011.

H75 Overview of H.264/AVC: http://www.csee.wvu.edu/~xinl/courses/ee569/H264_tutorial.pdf

Access to HM Software Manual: http://iphome.hhi.de/marpe/download/Performance_HEVC_VP9_X264_PCS_2013_preprint.pdf

H76 M.P. Krishnan, "Implementation and performance analysis of 2D order 16 integer transform in H.264 and AVS – China for HD video coding" M.S. Thesis, EE Dept., UT-Arlington, TX, Dec. 2010.

http://www.uta.edu/faculty/krrao/dip click on courses and then click on EE5359 Scroll down and go to Thesis/Project Title and click on M. Krishnan.

H77 B. Dey and M.K. Kundu, "Enhanced microblock features for dynamic background modeling in H.264/AVC video encoded at low bitrate", IEEE Trans. CSVT, (EARLY ACCESS).

H78 K. Minemura; K. Wong; R. Phan; K. Tanaka, "A Novel Sketch Attack for H.264/AVC Format-Compliant Encrypted Video," in *IEEE Trans. CSVT, (Early access)*. Conclusions are repeated here.

Conclusions: In this work, we proposed a novel sketch attack for H.264/AVC format-compliant encrypted video. Specifically, the macroblock bit stream size was exploited to sketch the outline of the original video frame directly from the encrypted video. In addition, the Canny edge map was considered as the ideal outline image and an edge similar score was modified for performance evaluation purposes. Experimental results suggest that the proposed sketch attack can extract visual information directly from the format-compliant encrypted video. Although the proposed and conventional methods can sketch the outline from the encrypted INTRA-frame, only the proposed MBS sketch attack method can sketch the outline from the encrypted INTER-frame, which outnumbers INTRA-frame, by far, in compressed video. Moreover, the proposed MBS sketch attack is verified to be more robust against compression when compared to the conventional sketch attacks. In view of this proposed sketch attack framework, we suggest that this framework should be considered for format compliant video encryption security analysis. In addition to determining the encryption modules in use by analyzing the encrypted video, we would like to extend this sketch attack framework to handle different video standards such as High Efficiency Video Coding (HEVC), Audio Video Standard (AVS) and Google VP9 as our future work.

Books on H.264

1. I. E. G. Richardson, "H.264 and MPEG-4 video compression: Video coding for Next Generation Multimedia," New York, Wiley, 2003.
2. S. Okubo et al., "H.264/AVC Textbook (in Japanese, title translated)," Tokyo, Impress Publisher, 2004.
3. S. Ono, T. Murakami and K. Asai, Ubiquitous technology, "Hybrid Video Coding – MPEG-4 and H.264 (In Japanese, title translated)", Tokyo, Ohmsha Press, 2005.
4. L. Chiariglione (Editor), "The MPEG Representation of Digital Media", Springer, 2012.

H.264 Standard, JM SOFTWARE

1. ITU – T and ISO/IEC JTC 1, "Advanced video coding for generic audiovisual services," ITU-T Rec. H.264 & ISO/IEC 14496-10, Version 1, May 2003; version 2, Jan 2004; Version 3 (with High family of profiles), Sept. 2004; version 4, July 2005 [online] Available: http://www.itu.int/rec/T-REC-H.264. JM 19.0 Reference software (6/19/2015) also updates on KTA.

DCT References

DCT1 N. Ahmed, M. Natarajan and K.R. Rao, "Discrete Cosine Transform," IEEE Trans. Computers, vol. C-23, pp. 90–93, Jan. 1974.

DCT2 W. Chen, C. H. Smith, and S. C. Fralick, "A fast computational algorithm for the discrete cosine transform," IEEE Trans. Commun., vol. COM-25, pp. 1004–1009, Sept. 1977.

DCT3 M. Vetterli and A. Ligtenberg "A discrete Fourier-cosine transform chip," IEEE Journal on Selected Areas of Communications, vol. 4, pp. 49–61, Jan. 1986.

DCT4 W.K. Cham and R.J. Clarke, "Application of the principle of dyadic symmetry to the generation of orthogonal transform," IEE Proc. F: Communications, Radar and Signal Processing, vol. 133, no. 3, pp. 264–270, June 1986.

DCT5 C. Loeffler, A. Lightenberg and G. Moschytz, "Practical fast 1-D DCT algorithms with 11 multiplications," Proc. IEEE ICASSP, vol. 2, pp. 988–991, Feb. 1989.

DCT6 W.K. Cham, "Development of integer cosine transforms by the principle of dyadic symmetry", IEE Proc. Communications, Speech and Vision, vol. 136, issue 4, pp. 276–282, 1989.

DCT7 M. Masera, M. Martina and G. Masera, "Adaptive approximated DCT architectures for HEVC", IEEE Trans. CSVT (Early access). This has several references related to integer DCT architectures. Same as [E381].

DCT8 S.-H. Bae, J. Kim and M. Kim, "HEVC-Based perceptually adaptive video coding using a DCT-Based local distortion detection probability", IEEE Trans. on Image Processing, vol. 25, pp. 3343–3357, July 2016. Same as [E377].

DCT9 Z. Zhang et al., "Focus and blurriness measure using reorganized DCT coefficients for auto focus" IEEE Trans. CSVT, Early access.

DCT10 S.-H. Bae and M. Kim, "DCT-QM; a dct-based quality degradation metric for image quality optimization problem", IEEE Trans. IP, vol. 25, pp. 4916–4930, Oct. 2016.

DCT11 M. Jridi and P.K. Meher, "A scalable approximate DCT architectures for efficient HEVC compliant video coding" IEEE Trans. CSVT, (early access) See also P.K. Meher et al., "Generalized architecture for integer DCT of lengths N = 8, 16 and 32", IEEE Trans. CSVT, vol. 24, pp. 168–178, Jan. 2014.

DCT12 V. Dhandapani and S. Ramachandran, "Area and power efficient DCT architecture for image compression", EURASIP Journal on Advances in Signal Processing, vol. 2014, 2014: 23.

DCT13 M. Narroschke, "Coding efficiency of the DCT and DST in hybrid video coding", IEEE J. on selected topics in signal processing, vol. 7, pp. 1062–1071, Dec. 2013.

DCT14 G. Fracastoro, S.M. Fosson and E. Magli, "Steerable discrete cosine transform", IEEE Trans. IP, vol. 26, pp. 303–314, Jan. 2017. Same as [E392].

DCT15 B. Zeng and J. Fu, "Directional discrete cosine transforms-a new framework for image coding," *IEEE Trans. CSVT,* vol. 18, pp. 305–313, March 2008.

DCT16 J.-S. Park et al., "2-D large inverse transform (16 × 16, 32 × 32) for HEVC (high efficiency video coding)", J. of Semiconductor Technology and Science, vol. 12, pp. 203–211, June 2012. Same as [E364].

DCT17 J. Dong, et al., "2-D order-16 integer transforms for HD video coding," IEEE Trans. CSVT, vol. 19, pp. 1462–1474, Oct. 2009.

DCT18 C.-K. Fong, Q. Han and W.-K. Cham, "Recursive Integer Cosine Transform for HEVC and Future Video Coding Standards" IEEE Trans. CSVT, vol. 27, pp. 326–336, Feb. 2017.

DCT19 C. Yeo et al., "Mode dependent transforms for coding directional intra prediction residuals", IEEE Trans. CSVT, vol. 22, pp. 545–554, April 2012.

DCT20 S.-H. Bae and M. Kim, "A DCT-based total JND profile for spatio-temporal and foveated masking effects", IEEE Trans. CSVT, vol. 27, pp. 1196–1207, June 2017.

DCT21 Z. Chen, Q. Han and W.-K. Cham, "Low-complexity order-64 integer cosine transform design and its applications in HEVC", IEEE Trans. CSVT (Under Review).

DCT22 J. Lee at al, "A compressed-domain corner detection method for a DCT-based compressed image", IEEE ICCE, Las Vegas, Jan. 2017.

DCT23 Z. Qian, W. Wang and T. Qiao, "An edge detection method in DCT domain", IEEE ICCE, Las Vegas, Jan. 2017.

DCT24 M. Wien, "High Efficiency Video Coding: Coding Tools and Specification", Springer, 2014. See Section 8.1.1 pages 206–210. This section explains how the integer DCTs adopted in HEVC are derived.

DCT25 P. Sjovall et al., "High-level synthesis implementation of HEVC 2-D DCT/DST on FPGA", IEEE ISCAS2017, pp. 1547–1551, New Orleans, Louisiana, March 2017.

DCT26 G. Pastuszak, "Hardware architectures for the H.265/HEVC discrete cosine transform," *IET Image Process.*, vol. 9, no. 6, pp. 468–477, 2015.

DCT27 R. Jeske, et al.,, "Low cost and high throughput multiplierless design of a 16 point 1-D DCT of the new HEVC video coding standard," *in Proc. Southern Conf. Programmable Logic*, Bento Goncalves, Spain, Mar. 2012.

DCT28 A. D. Darji and R. P. Makwana, "High-performance multiplierless DCT architecture for HEVC," *in Proc. Int. Symp. VLSI Design and Test*, Ahmedabad, India, June 2015.

DCT29 W. Zhao, T. Onoye, and T. Song, "High-performance multiplierless transform architecture for HEVC," *in Proc. IEEE ISCAS*, pp. 1668–1671, Beijing, China, May 2013.

DCT30 P. Arayacheeppreecha, S. Pumrin, and B. Supmonchai, "Flexible input transform architecture for HEVC encoder on FPGA," *in Proc. Int. Conf. Electrical Engineering/Electronics, Computer, Telecommunications and Information Tech.*, Hua Hin, Thailand, June 2015.

DCT31 G. Pastuszak and A. Abramowski, "Algorithm and architecture design of the H.265/HEVC intra encoder," *IEEE Trans. CSVT,* vol. 26, pp. 210–222, Jan. 2016.

DCT32 T. Ichita et al., "Directional discrete cosine transforms arising from discrete cosine and sine transforms for directional block-wise image representation", IEEE ICASSP2017, pp. 4536–4540, New Orleans, March 2017.

DCT33 G.C. Langelaar and R.L. Lagendik, "Optimal differential energy watermarking of DCT encoded images and video", IEEE Trans. Image Processing, vol. 10, pp. 148–158, Jan. 2001.

DCT-P1 Fong, Han and Cham [DCT18] have developed recursive integer cosine transform (RICT). This has orthogonal basis vectors and also has recursive property. Based on this, flow graph for order 32 RICT with flow graphs for orders 4, 8 and 16 embedded is shown in Figure 1. They also have implemented the RICT in HEVC HM13.0 and have demonstrated that the RICT has similar coding efficiency as the core transform in HEVC (see Tables VI thru XI). Using Figure 1, write down the sparse matrix factors (SMFs) for orders 4, 8, 16 and 32 RICTs.

DCT-P2 See DCT-P1. Using Table I and the SMFs for orders 4, 8, 16 and 32 RICTs write down the corresponding transform matrices. Verify the transform matrix for order 32 with that shown in Equation 16 of [DCT18].

DCT-P3 See DCT-P1. Forward and inverse architecture of order 4 RICT is shown in Figure 4. This structure is symmetric (Hardware of order 4 inverse RICT can be shared by the order 4 forward RICT.) Develop similar symmetric hardware structures for orders 8, 16 and 32.

DCT-P4 The authors in [DCT18] state that higher order RICTs such as orders 64 and 128 can be derived using their fast algorithms. Derive these.

DCT-P5 See DCT-P4. Draw flow graph for order 64 RICT with the flow graphs for orders 4, 8, 16 and 32 embedded similar to that shown in Figure 1.

DCT-P6 Repeat DCT-P5 for order 128 RICT.

DCT-P7 See DCT-P6. Extend the suggested values for $b_{N,i}$ for orders 64 and 128 RICTs (See Table I in [DCT18]).

DCT-P8 See DCT-P3. Develop similar symmetric hardware structures for orders 64 and 128 RICT.

DCT-P9 See Table V in [DCT18]. Extend the comparison to orders 64 and 128. This requires extending Loeffler's method to orders 64 and 128.

DCT-P10 See [DCT11] Jridi and Meher have developed approximate DCT architectures for efficient HEVC compliant video coding. Full-parallel and area constrained architectures for the proposed approximate DCT are described. Show that the equations given by (6a) and (6b) yield the 4-point DCT kernel given by Equation 4. Write down the SMFs based on Figure 1 for the 4-point integer 1-D DCT.

DCT-P11 See DCT-P10. Show that the equations given by (7)–(9) yield Equation 5 for the 8-point DCT kernel.

DCT-P12 See DCT-P11. Write down the SMFs for the 8-point integer 1-D DCT based on Figure 2. Show that these SMFs yield Equation 5 for the 8-point DCT kernel.

DCT-P13 See [DCT11]. Parallel architecture for approximate 8-point 1-D DCT is shown in Figure 2. Develop similar architecture for approximate 16-point 1-D DCT.

DCT-P15 See DCT-P13. Extend this to approximate 32-point 1-D DCT.

DCT-P16 See DCT-P14. Extend this to 32-point 1-D DCT.

DCT-P17 Chen, Han and Cham [DCT21] have developed a set of low complexity integer cosine transforms (LCICTs) and used them in HEVC HM13. These have fully factorizable structures and high circuit reusability. They conclude that order-64 ICTs can significantly improve the coding performance under high QP (low bitrate) which makes them beneficial in low bitrate applications such as video conferencing and video surveillance. Using Equations 5–10, write down the LCITC matrices for orders 8, 16, 32 and 64.

DCT-P18 See DCT-P17. Using Figure 3 (flow graph) write down the SMFs for LCICTs of order 8, 16, and 32.

DCT-P19 See DCT-P17. Extend the flow graph for LCICT of order 64.

DCT-P20 See DCT-P17. Write down the SMFs for LCICT of order 64.

DCT-P21 See [DCT21]. In Table IV the BD bitrates for LCICT are compared with those for another order 64 ICT (see reference 14 at the end) using various test sequences under AI, RA, and LDB. Confirm these results.

DCT-P22 The flow graph for order-32 LCICT (flow graphs for order-8 and 16 LCICT are embedded) is shown in Figure 1 [DCT21]. Draw the flow graph for the inverse LCICTs.

DCT-P23 Based on DCT-P22, write down the corresponding SMFs for these inverse LCICTs.

DCT-P24 See DCT-P22 and DCT-P23. Evaluate the orthogonality of these LCICTs. Hint: matrix multiply the forward LCICT with the matrix for corresponding inverse LCICT. Are these matrices truly orthogonal?

DCT-P25 M. Wien, "High Efficiency Video Coding: Coding Tools and Specification", Springer, 2014. See Section 7.3.3 Determination of the interpolation filter (IF) coefficients pages 194–196 for the theory behind how these IF coefficients are derived using DCT. Derive these IF coefficients independently and confirm that they match with the coefficients listed in Tables 5.4 and 5.5. Plot the corresponding transfer functions (See Figures 7.16 and 7.17 – pages 201–202).

DCT-P26 Sjovall et al. [DCT25] have developed a high level synthesis implementation of HEVC 2-D DCT/DST (8/16/32 point DCT units and 4-point DCT/DST unit) on FPGA. Develop this architecture and verify the results shown in Tables 1 and 2.

DCT-P27 See DCT-P26. Extend the high level synthesis to 2D-64 point DCT.

MPA.P8: Transform Coding (IEEE ICIP 2016)

Session Type: Poster
Time: Monday, September 26, 14:00–15:20
Location: Room 301 CD: Poster Area 8
Session Chair: Onur Guleryuz, LG Electronics Mobile Research Lab

MPA.P8.1: A Staircase Transform Coding Scheme for Screen Content Video Coding

Cheng Chen; *University of Iowa*
Jingning Han; *Google Inc.*
Yaowu Xu; *Google Inc.*
James Bankoski; *Google Inc.*

MPA.P8.2: Fast MCT Optimization for the Compression of Whole-Slide Images

Miguel Hernandez-Cabronero; *University of Warwick*

Victor Sanchez; *University of Warwick*
Francesc Auli-LLinas; *Universitat Autònoma de Barcelona*
Joan Serra-Sagristá; *Universitat Autònoma de Barcelona*

MPA.P8.3: GBST: Separable Transforms Based Online Graphs for Predictive Video Coding

Hilmi E. Egilmez; *University of Southern California*
Yung-Hsuan Chao; *University of Southern California*
Antonio Ortega; *University of Southern California*
Bumshik Lee; *LG Electronics Inc.*
Sehoon Yea; *LG Electronics Inc.*

MPA.P8.4: H.264 Intra Coding with Transforms Based on Prediction Inaccuracy Modeling

Xun Cai; *Massachusetts Institute of Technology*
Jae Lim; *Massachusetts Institute of Technology*

MPA.P8.5: Row-Column Transforms: Low-Complexity Approximation of Optimal Non-Separable Transforms

Hilmi E. Egilmez; *University of Southern California*
Onur G. Guleryuz; *LG Electronics Inc.*
Jana Ehmann; *LG Electronics Inc.*
Sehoon Yea; *LG Electronics Inc.*

MPA.P8.6: Extended Block-Lifting-Based Lapped Transforms

Taizo Suzuki; *University of Tsukuba*
Hiroyuki Kudo; *University of Tsukuba*

MPA.P8.7: Transform-Coded PEL-Recursive Video Compression

Jana Ehmann; *LG Electronics Inc.*
Onur G. Guleryuz; *LG Electronics Inc.*
Sehoon Yea; *LG Electronics Inc.*

MPA.P8.8: Rate-Constrained Successive Elimination of Hadamard-Based SATDS

Ismael Seidel; *Federal University of Santa Catarina*
Luiz Cancellier; *Federal University of Santa Catarina*
José Luís Güntzel; *Federal University of Santa Catarina*
Luciano Agostini; *Federal University of Pelotas*

HEVC (High Efficiency Video Coding) References

E3. K.H. Lee *et al.*, "Technical considerations for Ad Hoc Group on new challenges in video coding standardization," ISO/IEC MPEG 85th meeting, M15580, Hanover, Germany, July 2008.

E4. E. Alshina *et al.*, "Technical considerations on new challenges in video coding standardization," ISO/IEC MPEG 86th meeting, M15899, Busan, South Korea, Oct. 2008.

E5. Y. Ye and M. Karczewicz, "Improved H.264 intra coding based on bi-directional intra prediction, directional transform, and adaptive coefficient scanning," IEEE Int'l Conf. Image Process.'08 (ICIP08), pp. 2116–2119, San Diego, CA, Oct. 2008.

E6. G.J. Sullivan, "The high efficiency video coding (HEVC) standardization initiative," Power Point slides, 7th June 2010.

E7. G.J. Sullivan and J.-R. Ohm, "Recent developments in standardization of high efficiency video coding (HEVC)," Proc. SPIE, Applications of Digital Image Processing XXXIII, vol. 7798, pp. 77980V-1 through V-7, San Diego, CA, 1–3 Aug. 2010.

E8. R. Joshi, Y.A. Reznik and M. Karczewicz, "Efficient large size transforms for high-performance video coding," Proc. SPIE, vol. 7798, San Diego, CA, Aug. 2010.

E9. IEEE Trans. on CSVT, vol. 20, Special section on HEVC (several papers), Dec. 2010.

E10. M. Karczewicz et al., "A hybrid video coder based on extended macroblock sizes, improved interpolation and flexible motion representation," IEEE Trans. CSVT, vol. 20, pp. 1698–1708, Dec. 2010.

E11. T. Wiegand, B. Bross, W.-J. Han, J.-R. Ohm, and G. J. Sullivan, WD3: Working Draft 3 of High-Efficiency Video Coding, Joint Collaborative Team on emerging HEVC standard on Video Coding (JCT-VC) of ITU-T VQEG and ISO/IEC MPEG, Doc. JCTVC-E603, Geneva, CH, March 2011.

E12. S. Jeong et al., "High efficiency video coding for entertainment quality," ETRI Journal, vol. 33, pp. 145–154, April 2011.

E13. K. Asai et al., "New video coding scheme optimized for high-resolution video sources," IEEE Journal of Selected Topics in Signal Processing, vol. 5, no. 7, pp. 1290–1297, Nov. 2011.

E14. IEEE Journal of Selected Topics in Signal Processing, vol. 5, no. 7, Nov. 2011 (several papers on HEVC), Introduction to the issue on emerging technologies for video compression.

E15. Z. Ma, H. Hu, and Y. Wang, "On complexity modeling of H.264/AVC video decoding and its application for energy efficient decoding," IEEE Trans. Multimedia, vol. 13, no. 6, pp. 1240–1255, Nov. 2011.

E16. J. Dong and K.N. Ngan, "Adaptive pre-interpolation filter for high efficiency video coding ", J. VCIR, vol. 22, pp. 697–703, Nov. 2011.

E17. W. Ding et al., "Fast mode dependent directional transform via butterfly-style transform and integer lifting steps", J. VCIR, vol. 22, pp. 721–726, Nov. 2011.

E18. F. De Simone et al., "Towards high efficiency video coding; subjective evaluation of potential coding technologies", J. VCIR, vol. 22, pp. 734–748, Nov. 2011.

E19. B. Bross et al., "High efficiency video coding (HEVC) text specification draft 6," Joint Collaborative Team on Video Coding (JCT-VC) of ITU-T SG16 WP3 and ISO/IEC JTC1/SC29/WG11, 7th Meeting: Geneva, CH, 21–30 Nov. 2011.

E20. Z. Ma and A. Segall, "Low resolution decoding for high-efficiency video coding," IASTED SIP-2011, Dallas, TX, Dec. 2011.

E21. Y.-J. Yoon et al., "Adaptive prediction block filter for video coding", ETRI Journal, vol. 34, pp. 106–109, Feb. 2012. (See the references) http://etrij.etri.re.kr.

E22. B. Li, G.J. Sullivan and J. Xu, "Compression performance of high efficiency video coding (HEVC) working draft 4", IEEE ISCAS, pp. 886–889, session B1L-H, Seoul, Korea, May 2012.

E23. N. Ling, "High efficiency video coding and its 3D extension: A research perspective," Keynote Speech, ICIEA, Singapore, July 2012.

E24. C. Fogg, "Suggested figures for the HEVC specification", ITU-T & ISO/IEC JCTVC-J0292r1, July 2012.

E25. J.-R, Ohm, G. J. Sullivan, and T. Wiegand, "Developments and trends in 2D and 3D video coding standardization: HEVC and more," IEEE ICIP 2012, Tutorial, Orlando, FL, 30 Sept. 2012.

E26. A. Saxena and F.C. Fernandes, "On secondary transforms for prediction residual", IEEE ICIP 2012, Orlando, FL, pp. 2489–2492, Sept. 2012.
E27. "High efficiency video coding" MA.P2, IEEE ICIP 2012, Poster session, (several papers), Orlando, FL, Sept.–Oct., 2012. These are listed below (E28–E40).
E28. S. Ma, J. Si and S. Wang, "A study on the rate distortion modeling for high efficiency video coding", IEEE ICIP 2012, pp. 181–184, Orlando, FL, Oct. 2012.
E29. A. Gabriellini et al., "Spatial transform skip in the emerging high efficiency video coding standard", IEEE ICIP 2012, pp. 185–188, Orlando, FL, Oct. 2012.
E30. H. Vianna et al., "High performance hardware architectures for the inverse rotational transform of the emerging HEVC standard", IEEE ICIP 2012, pp. 189–192, Orlando, FL, Oct. 2012.
E31. H. Aoki, "Efficient quantization parameter coding based on intra/inter prediction for visual quality conscious video coders", IEEE ICIP 2012, pp. 193–196, Orlando, FL, Oct. 2012.
E32. X. Zhang et al., "New chroma intra prediction modes based on linear model for HEVC", IEEE ICIP 2012, pp. 197–200, Orlando, FL, Oct. 2012.
E33. D. Palomino, "A memory aware and multiplierless VLSI architecture for the complete intra prediction of the HEVC emerging standard", IEEE ICIP 2012, pp. 201–204, Orlando, FL, Oct. 2012.
E34. H. Schwarz, "Extension of high efficiency video coding (HEVC) for multiview video and depth data", IEEE ICIP 2012, pp. 205–208, Orlando, FL, Oct. 2012.
E35. M. Budagavi and V. Sze, "Unified forward + inverse transform architecture for HEVC", IEEE ICIP 2012, pp. 209–212, Orlando, FL, Oct. 2012.
E36. C.C. Chi, "Improving the parallelization efficiency of HEVC decoding", IEEE ICIP 2012, pp. 213–216, Orlando, FL, Oct. 2012.
E37. G. Correa et al., "Motion compensated tree depth limitation for complexity control of HEVC encoding", IEEE ICIP 2012, pp. 217–220, Orlando, FL, Oct. 2012.
E38. M. Zhang, C. Zhao and J. Xu, "An adaptive fast intra mode decision in HEVC", IEEE ICIP 2012, pp. 221–224, Orlando, FL, Oct. 2012.

E39. J. Stankowski et al., "Extensions of the HEVC technology for efficient multiview video coding", IEEE ICIP 2012, pp. 225–228, Orlando, FL, Oct. 2012.

E40. L. Kerofsky, K. Misra and A. Segall, "On transform dynamic range in high efficiency video coding", IEEE ICIP 2012, pp. 229–232, Orlando, FL, Oct. 2012.

E41. M.E. Sinangil et al., "Memory cost vs. coding efficiency trade-offs for HEVC motion estimation engine," IEEE ICIP 2012, pp. 1533–1536, Orlando, FL, Sept.–Oct., 2012.

E42. M.E. Sinangil et al., "Hardware – aware motion estimation search algorithm development for HEVC standard", IEEE ICIP 2012, pp. 1529–1532, Orlando, FL, Sept.–Oct., 2012.

E43. M. Zhang, C. Zhao and J. Xu, "An adaptive fast intra mode decision in HEVC", IEEE ICIP 2012, pp. 221–224, Orlando, FL, Sept.–Oct., 2012.

E44. H. Zhang and Z. Ma, "Fast intra prediction for high efficiency video coding", Pacific Rim Conf. on Multimedia, PCM2012, Singapore, Dec. 2012. See [E147].

E45. Special issue on emerging research and standards in next generation video coding, IEEE Trans. CSVT, vol. 22, pp. 1646–1909, Dec. 2012.

E46. B. Shrestha, "Investigation of image quality of Dirac, H.264 and H.265," Report for EE 5359: Multimedia Processing, EE Dept., UTA, Arlington, TX, 2012.
http://www.uta.edu/faculty/krrao/dip/

E47. B. Li, G.J. Sullivan, and J. Xu, "Compression performance of high efficiency video coding working draft 4," IEEE ISCAS 2012, Session on HEVC, pp. 886–889, Seoul, Korea, May 2012.

E48. http://www.h265.net has info on developments in HEVC NGVC – Next generation video coding.

E49. JVT KTA reference software http://iphome.hhi.de/suehring/tml/download/KTA/

E50. JM11KTA2.3 (http://www.h265.net/2009/04/kta-software-jm11kta23.html).

E51. Permanent Link: Mode-Dependent Directional Transform (MDDT) in JM/KTA
TMUC HEVC Software – http://hevc.kw.bbc.co.uk/svn/jctvc-tmuc/.

E52. F. Bossen, D. Flynn and K. Shring (July 2011), "HEVC reference software manual" Online available: http://phenix.int-evry.fr/jct/doc_end_user/documents/6_Torino/wg11/JCTVC-F634-v2.zip

E53. http://phenix.it-sudparis.eu/jct/doc_end_user/documents/7_Geneva/wg11/JCTVC-G399-v3.zip
E54. http://www.itu.int/ITU-T/studygroups/com16/jct-vc/index.html.
E55. JCT-VC documents are publicly available at http://ftp3.itu.ch/av-arch/jctvc-site and http://phenix.it-sudparis.eu/jct/.
E56. HEVC Model 9.0, available for download at https://hevc.hhi.fraunhofer.de/svn/svn_HEVCSoftware/tags/HM-9.0/
The latest version available at the time this chapter was prepared is HM-9.0.
Version HM 8.0 is available at: http://hevc.kw.bbc.co.uk/trac/browser/tags/HM-8.0
E57. J. Nightingale, Q. Wang and C. Grecos, "HEVStream; A framework for streaming and evaluation of high efficiency video coding (HEVC) content in loss-prone networks", IEEE Trans. Consumer Electronics, vol. 59, pp. 404–412, May 2012.
E58. D. Marpe et al., "Improved video compression technology and the emerging high efficiency video coding standard", IEEE International Conf. on Consumer Electronics, pp. 52–56, Berlin, Germany, Sept. 2011.
E59. HM9 High efficiency video coding (HEVC) Test Model 9 (HM 9) encoder description JCTVC-K1002v2, Shanghai meeting, Oct. 2012.
E60. B. Bross et al., "High efficiency video coding (HEVC) text specification draft 8", JCTVC-J1003, July 2012. http://phenix.int-evry.fr/jct/doc_end_user/current_document.Php?id=5889
E61. G.J. Sullivan et al., "Overview of the high efficiency video coding (HEVC) standard", IEEE Trans. CSVT, vol. 22, pp. 1649–1668, Dec. 2012.
E62. J.-R. Ohm et al., "Comparison of the coding efficiency of video coding standards – including high efficiency video coding (HEVC)", IEEE Trans. CSVT, vol. 22, pp. 1669–1684, Dec. 2012. Software and data for reproducing selected results can be found at ftp://ftp.hhi.de/ieee-tcsvt/2012
E63. F. Bossen et al., "HEVC complexity and implementation analysis", IEEE Trans. CSVT, vol. 22, pp. 1685–1696, Dec. 2012.
E64. B. Bross et al., "Block merging for quadtree based partitioning in HEVC", SPIE Applications of digital image processing XXXV, vol. 8499, paper 8499-26, Aug. 2012.
E65. P. Hanhart et al., "Subjective quality evaluation of the upcoming HEVC video compression standard", SPIE Applications of digital image processing XXXV, vol. 8499, paper 8499–30, Aug. 2012.

E66. M. Horowitz et al., " Informal subjective quality comparison of video compression performance of the HEVC and H.264/MPEG-4 AVC standards for low delay applications", SPIE Applications of digital image processing XXXV, vol. 8499, paper 8499-31, (W84990-1), Aug. 2012.

E67. V. Sze and M. Budagavi, "High throughput CABAC entropy coding in HEVC", IEEE Trans. CSVT, vol. 22, pp. 1778–1791, Dec. 2012.

E68. V. Sze and A.P. Chandrakasan, "Joint algorithm-architecture optimization of CABAC to increase speed and reduce area cost", IEEE ISCAS, Proc. pp. 1577–1580, 2011.

E69. H. Schwartz, D. Marpe and T. Wiegand, "Overview of the scalable video coding extension of the H.264/AVC standard", IEEE Trans. CSVT, vol. 17, pp. 1103–1120, Sept. 2007. This is a special issue on scalable video coding in H.264/AVC.0

E70. JSVM – joint scalable video model reference software for scalable video coding. (online) http://ip.hhi.de/imagecom_GI/savce/downloads/SVC-Reference-software.htm

E71. H. Lakshman, H. Schwarz and T. Wiegand, "Generalized interpolation based fractional sample motion compensation", IEEE Trans. CSVT, vol. 23, pp. 455–466, March 2013. A sample C++ implementation (that can be tested using the first test model of HEVC, HM1.0) can be found in http://phenix.int-evry.fr/jct/doc_end_user/documents/4_Daegu/wg11/JCTVC-D056-v2.zip

E72. K. Ugur et al., "High performance low complexity video coding and the emerging high efficiency video coding standard", IEEE Trans. CSVT, vol. 20, pp. 1688–1697, Dec. 2010.

E73. A. Norkin et al., "HEVC deblocking filter", IEEE Trans. CSVT, vol. 22, pp. 1746–1754, Dec. 2012. Corrections to "HEVC deblocking filter", IEEE Trans. CSVT, vol. 23, pp. 2141, Dec. 2013.

E74. J. Sole et al., "Transform coefficient coding in HEVC", IEEE Trans. CSVT, vol. 22, pp. 1765–1777, Dec. 2012.

E75. JSVM9 Joint scalable video model 9 http://ip.hhi.de/imagecom_GI/savce/downloads

E76. Z. Shi, X. Sun and F. Wu, "Spatially scalable video coding for HEVC", IEEE Trans CSVT, vol. 22, pp. 1813–1826, Dec. 2012.

E77. A. Krutz et al., "Adaptive global motion temporal filtering for high efficiency video coding", IEEE Trans. CSVT, vol. 22, pp. 1802–1812, Dec. 2012.

E78. J. Lainema et al., "Intra coding of the HEVC standard", IEEE Trans. CSVT, vol. 22, pp. 1792–1801, Dec. 2012.

E79. H. Li, B. Li and J. Xu, "Rate distortion optimized reference picture management for high efficiency video coding", IEEE Trans. CSVT, vol. 22, pp. 1844–1857, Dec. 2012.

E80. Special issue on video coding: HEVC and beyond, IEEE Journal of selected topics in signal processing, vol. 7, Dec. 2013.

E81. G. Bjontegaard, "Calculation of average PSNR differences between RD-Curves", ITU-T SG16, Doc. VCEG-M33, 13th VCEG meeting, Austin, TX, April 2001.http://wfpt3.itu.int/av-arch/video-site/0104_Aus/VCEG-M33.doc

A graphical explanation of BD-PSNR and BS-bitrate is shown in [E202]

E82. G. Bjontegaard, "Improvements of the BD-PSNR model", ITU-T SG16 Q.6, Doc. VCEG-AI11, Berlin, Germany, July 2008.

E83. T. Schierl et al., "System layer integration of HEVC", IEEE Trans. CSVT, vol. 22, pp. 1871–1884, Dec. 2012.

E84. G. Correa et al., "Performance and computational complexity assessment of high efficiency video encoders", IEEE Trans. CSVT, vol. 22, pp. 1899–1909, Dec. 2012.

E85. M. Zhou et al., "HEVC lossless coding and improvements", IEEE Trans. CSVT, vol. 22, pp. 1839–1843, Dec. 2012.

E86. C. Yeo, Y. H. Tan and Z. Li, "Dynamic range analysis in high efficiency video coding residual coding and reconstruction", IEEE Trans. CSVT, vol. 23, pp. 1131–1136, July 2013.

E87. C.-M. Fu et al., "Sample adaptive offset in the HEVC standard", IEEE Trans. CSVT, vol. 22, pp. 1755–1764, Dec. 2012.

E88. C.C. Chi et al., "Parallel scalability and efficiency of HEVC parallelization approaches", IEEE Trans. CSVT, vol. 22, pp. 1827–1838, Dec. 2012.

E89. J. Vanne et al., "Comparative rate-distortion complexity analysis of HEVC and AVC video codecs", IEEE Trans. CSVT, vol. 22, pp. 1885–1898, Dec. 2012.

E90. Y. Yuan et al., "Quadtree based non-square block structure for interframe coding in high efficiency video coding", IEEE Trans. CSVT, vol. 22, pp. 1707–1719, Dec. 2012.

E91. R. Sjoberg et al., "Overview of high-level syntax and reference picture management", IEEE Trans. CSVT, vol. 22, pp. 1858–1870, Dec. 2012.

E92. P. Helle et al., "Block merging for quadtree-based partitioning in HEVC", IEEE Trans. CSVT, vol. 22, pp. 1720–1731, Dec. 2012.

E93. I.-K. Kim et al., "Block partitioning structure in HEVC standard", IEEE Trans. CSVT, vol. 22, pp. 1697–1706, Dec. 2012.

E94. T. Lin et al., "Mixed chroma sampling-rate high efficiency video coding for full-chroma screen content", IEEE Trans. CSVT, vol. 23, pp. 173–185, Jan. 2013.

E95. E. Peixoto and E. Izquierdo, "A new complexity scalable transcoder from H.264/AVC to the new HEVC codec", IEEE ICIP 2012, session MP.P1.13 (Poster), Orlando, FL, Sept.–Oct. 2012.

E96. K. Anderson, R. Sjobetg and A. Norkin, "BD measurements based on MOS", (online), ITU-T Q6/SG16, document VCEG-AL23, Geneva, Switzerland, July 2009. Available: http://wfpt3.itu.int/av-arch/video-site/0906_LG/VCEG-AL23.zip

E97. HEVC open source software (encoder/decoder)
https://hevc.hhi.fraunhofer.de/svn/svn_HEVCSoftware/tags/HM-10.1
JCT-VC (2014) Subversion repository for the HEVC test model reference software
https://hevc.hhi.fraunhofer.de/svn/svn_HEVCSoftware/
http://hevc.kw.bbc.co.uk/svn/jctvc-al24 (mirror site)

E98. T. Shanableh, E. Peixoto and E. Izquierdo, "MPEG-2 to HEVC transcoding with content based modeling", IEEE Trans. CSVT, vol. 23, pp. 1191–1196, July 2013.
References [E97] thru [E101] are from VCIP 2012. Pl access http://www.vcip2012.org. There are also some papers on HEVC in the Demo session.

E99. G.J. Sullivan, "HEVC: The next generation in video compression". Keynote speech, Visual Communications and Image Processing, VCIP 2012, San Diego, CA, 27–30 Nov. 2012.

E100. Two oral sessions on HEVC (W03 and F01). Also panel discussion on "All about HEVC" Moderator: Shipeng Li, Microsoft Research Asia.

E101. X. Zhang, S. Liu and S. Lei, "Intra mode coding in HEVC standard", Visual Communications and Image Processing, VCIP 2012, pp. 1–6, San Diego, CA, 27–30, Nov. 2012.

E102. L. Yan et al., "Implementation of HEVC decoder on X86 processors with SIMD optimization", Visual Communications and Image Processing, VCIP 2012, San Diego, CA, 27–30, pp. 1–6, Nov. 2012.

E103. W. Dai and H. Xiong, "Optimal intra coding of HEVC by structured set prediction mode with discriminative learning", Visual Communications and Image Processing, VCIP 2012, pp. 1–6, San Diego, CA, 27–30, Nov. 2012.

E104. G. Li et al., "Uniform probability model for deriving intra prediction angular table in high efficiency video coding", Visual Communications and Image Processing, VCIP 2012, pp. 1–5, San Diego, CA, 27–30, Nov. 2012.

E105. Y. Duan, "An optimized real time multi-thread HEVC decoder", Visual Communications and Image Processing, VCIP 2012, pp. 1–1, San Diego, CA, 27–30, Nov. 2012.
References [E99] thru [E105] are from VCIP 2012. Pl access http://www.vcip2012.org. There are also some papers on HEVC in the Demo session.

E106. S. Gangavathi, "Complexity reduction of H.264 using parallel programming", M.S. Thesis, EE Dept., UTA, Arlington, TX. Dec. 2012.

E107. M.T. Pourazad et al., "HEVC: The new gold standard for video compression", IEEE CE magazine, vol. 1, issue 3, pp. 36–46, July 2012.

E108. Y. Zhang, Z. Li and B. Li, "Gradient-based fast decision for intra prediction in HEVC", Visual Communications and Image Processing, VCIP 2012, San Diego, CA, 27–30, Nov. 2012.

E109. H. Lv et al., "A comparison of fractional-pel interpolation filters in HEVC and H.264/AVC", Visual Communications and Image Processing, VCIP 2012, San Diego, CA, 27–30, Nov. 2012. Several references leading to the selection of DCTIF are listed at the end.

E110. J. Wang et al., "Multiple sign bits hiding for high efficiency video coding", Visual Communications and Image Processing, VCIP 2012, San Diego, CA, 27–30, Nov. 2012.

E111. C.-M. Fu et al., "Sample adaptive offset for HEVC", in Proc. Intl. workshop on multimedia signal processing, (MMSP), Hangzhou, China, Oct. 2011.

E112. S. Subbarayappa, "Implementation and Analysis of Directional Discrete Cosine Transform in Baseline Profile in H.264", M.S. Thesis, EE Dept., UTA, Arlington, Texas, 2012. http://www.uta.edu/faculty/krrao/dip/

E113. P. Anjanappa, "Performance analysis and implementation of mode dependent DCT/DST in H.264/AVC", M.S. Thesis, EE Dept., UTA, Arlington, Texas, 2012. http://www.uta.edu/faculty/krrao/dip/

E114. Q. Cai et al., "Lossy and lossless intra coding performance evaluation: HEVC, H.264/AVS, JPEG 2000 and JPEG LS", APSIPA, Los Angeles, CA, Dec. 2012.

E115. K. Chen et al., "Efficient SIMD optimization of HEVC encoder over X86 processors", APSIPA, Los Angeles, CA, Dec. 2012.

E116. H. Lv et al., "An efficient NEON-based quarter-pel interpolation method for HEVC", APSIPA, Los Angeles, CA, Dec. 2012.

E117. Q. Yu et al., "Early termination of coding unit splitting for HEVC", APSIPA, Los Angeles, CA, Dec. 2012.

E118. J.M. Nightingale, Q. Wang and C. Grecos, "Priority-based methods for reducing the input of packet loss on HEVC encoded video streams", SPIE/EI, 8656-15, Feb. 2013.

E119. Joint preliminary call for proposals on scalable video coding extensions of high efficiency video coding (HEVC), Geneva, Switzerland, May 2012.

E120. S. Kamp and M. Wien, "Decoder-side motion vector derivation for block-based video coding", IEEE Trans. CSVT, vol. 22, pp. 1732–1745, Dec. 2012.

E121. A. Saxena, F. Fernandes and Y. Reznik, "Fast transforms for Intra-prediction-based image and video coding," in Proc. IEEE Data Compression Conference (DCC'13), Snowbird, UT, March 20–23, 2013.

E122. R. Chivukula and Y. Reznik, "Fast computing of discrete cosine and sine transforms of types VI and VII," in Applications of Digital Image Processing XXXIV, Proc. SPIE, A. G. Tescher, Ed., vol. 8135, pp. 813505–1–14, 2011.

E123. M. Zhou et al., "HEVC lossless coding and improvements", SPIE/EI, vol. 8666-10, Burlingame, CA, Feb. 2013.

E124. R. Garcia and H. Kalva, "Subjective evaluation of HEVC in mobile devices", SPIE/EI, vol. 8667-19, Burlingame, CA, Feb. 2013. See also "Human mobile-device interaction on HEVC and H.264 subjective evaluation for video use in mobile environment", IEEE ICCE, pp. 639–640, Jan. 2013. See [E210].

E125. B. Bross et al., "High efficiency video coding (HEVC) text specification draft 10 (for FDIS & Consent)", JCTVC-L1003v13, Jan. 2013.

E126. X. Xiu et al., "Improved motion prediction for scalable extensions of HEVC", SPIE/EI, vol. 8666-5, Burlingame, CA, Feb. 2013.

E127. T. Hinz et al., "An HEVC extension for spatial and quality scalable video coding", SPIE/EI, vol. 8666-6, Burlingame, CA, Feb. 2013.
E128. K. Misra et al., "Scalable extensions of HEVC for next generation services", SPIE/EI, vol. 8666-7, Burlingame, CA, Feb. 2013.
E129. ISO/IEC JTC-1/SC29/WG11 w12957, "Joint call for proposals on scalable video coding extensions of high efficiency video coding", July 2012.
E130. S.G. Deshpande et al., "An improved hypothetical reference decoder for HEVC", SPIE/EI, vol. 8666-9, Burlingame, CA, Feb. 2013.
E131. J. Han et al., "Towards jointly optimal spatial prediction and adaptive transform in video/image coding", IEEE ICASSP, pp. 726–729, Mar. 2010.
E132. A. Saxena and F. Fernandes, "CE7: Mode-dependent DCT/DST for intra prediction in video coding", ITU-T/ISO-IEC Document: JCTVC-D033, Jan. 2011.
E133. A. Saxena and F. Fernandes, "Mode dependent DCT/DST for intra prediction in block-based image/video coding", IEEE ICIP, pp. 1685–1688, Sept. 2011.
E134. Y. Kim et al., "A fast intra-prediction method in HEVC using rate-distortion estimation based on Hadamard transform", ETRI Journal, vol. 35, #2, pp. 270–280, April 2013.
E135. D. Flynn, G. Martin-Cocher and D. He, "Decoding a 10-bit sequence using an 8-bit decoder", JCTVC-M0255, April, 2013.
E136. J.-R. Ohm, T. Wiegand and G. J. Sullivan, "Video coding progress; The high efficiency video coding (HEVC) standard and its future extensions", IEEE ICASSP, Tutorial, Vancouver, Canada, June 2013.
E137. D. Springer et al., "Robust rotational motion estimation for efficient HEVC compression of 2D and 3D navigation video sequences", IEEE ICASSP 2013, pp. 1379–1383, Vancouver, Canada, 2013.
E138. H. Li, Y. Zhang and H. Chao, "An optimally scalable and cost-effective fractional motion estimation algorithm for HEVC", IEEE ICASSP 2013, pp. 1399–1403, Vancouver, Canada, June 2013.
E139. Several papers on HEVC in the poster session IVMSP-P3: Video coding II, IEEE ICASSP 2013, Vancouver, Canada, June 2013.
E140. Z. Pan et al., "Early termination for TZSearch in HEVC motion estimation", IEEE ICASSP 2013, pp. 1389–1392, June 2013.
E141. D.-K. Kwon, M. Budagavi and M. Zhou, "Multi-loop scalable video codec based on high efficiency video coding (HEVC)", IEEE ICASSP 2013, pp. 1749–1753, June 2013.

E142. H.R. Tohidpour, M.T. Pourazad and P. Nasiopoulos, "Content adaptive complexity reduction scheme for quality/fidelity scalable HEVC", IEEE ICASSP 2013, pp. 1744–1748, June 2013.

E143. Y.H. Tan, C. Yeo and Z. Li, "Residual DPCM for lossless coding in HEVC", IEEE ICASSP 2013, pp. 2021–2025, June 2013.

E144. M. Wien, "HEVC – coding tools and specifications", Tutorial, IEEE ICME, San Jose, CA, July 2013.

E145. T.L. da Silva, L.V. Agostini and L.A. da Silva Cruz, "HEVC intra coding acceleration based on tree intra-level mode correlation", IEEE Conference 2013 Signal Processing: Algorithms, Architectures, Arrangements, and Applications (SPA 2013), Poznan, Poland, Sept. 2013.

E146. E. Peixoto et al., "An H.264/AVC to HEVC video transcoder based on mode mapping", IEEE ICIP 2013, WP.P4.3 (poster), Melbourne, Australia, 15–18 Sept. 2013.

E147. E. Wige et al., "Pixel-based averaging predictor for HEVC lossless coding", IEEE ICIP 2013,WP.P311 (poster), Melbourne, Australia, 15–18 Sept. 2013.

E148. E. Peixoto, T. Shanableh and E. Izquierdo, "H.264/AVC to HEVC video transcoder based on dynamic thresholding and content modeling", IEEE Trans. on CSVT, vol. 24, pp. 99–112, Jan. 2014.

E149. H. Zhang and Z. Ma, "Fast intra mode decision for high-efficiency video coding", IEEE Trans. CSVT, Vol. 24, pp. 660–668, April 2014. See [E42].

E150. Y. Tew and K.S. Wong, "An overview of information hiding in H.264/AVC compressed video", IEEE Trans. CSVT, vol. 24, pp. 305–319, Feb. 2014. (repeated see E175).

E151. M. Zhou, V. Sze and M. Budagavi, "Parallel tools in HEVC for high-throughput processing", Applications of digital image processing, SPIE, vol. 8499, pp. 849910-1 thru 849910-12, 2012.

E152. W.-H. Peng and C.-C. Chen, "An interframe prediction technique combining template matching and block motion compensation for high efficiency video coding", IEEE Trans. CSVT, vol. 23, pp. 1432–1446, Aug. 2013.

E153. S. Blasi, E. Peixoto and E. Izqueirdo, "Enhanced Inter – Prediction using merge prediction transformation in the HEVC codec," IEEE ICASSP 2013, pp. 1709–1713, Van Couver, CA, June 2013.

E154. L.-L. Wang and W.-C. Siu, "Novel Adaptive Algorithm for Intra Prediction With Compromised Modes Skipping and Signaling Processes in HEVC", IEEE Trans. CSVT, Vol. 23, pp. 1686–1694, Oct. 2013.

E155. D. Flynn, J. Sole and T. Suzuki, "High efficiency video coding (HEVC) range extensions text specification", Draft 4, JCT-VC. Retrieved 2013-08-07.

E156. M. Budagavi and D.-Y. Kwon, "Intra motion compensation and entropy coding improvements for HEVC screen content coding", IEEE PCS, pp. 365–368, San Jose, CA, Dec. 2013.

E157. M. Naccari et al., "Improving inter prediction in HEVC with residual DPCM for lossless screen content coding", IEEE PCS, pp. 361–364, San Jose, CA, Dec. 2013.

E158. E. Alshina et al., "Inter-layer filtering for scalable extension of HEVC", IEEE PCS, pp. 382–385, San Jose, CA, Dec. 2013.

E159. J. Zhao, K. Misra and A. Segall, "Content-adaptive upsampling for scalable video coding", IEEE PCS, pp. 378–381, San Jose, CA, Dec. 2013.

E160. G.J. Sullivan et al., "Standardized extensions of high efficiency video coding (HEVC)", IEEE Journal of selected topics in signal processing, vol. 7, pp. 1001–1016, Dec. 2013.

E161. F. Pescador et al., "On an Implementation of HEVC Video Decoders with DSP Technology", IEEE ICCE, pp. 121–122, Las Vegas, NV, Jan. 2013.

E162. F. Pescador et al., "A DSP HEVC decoder implementation based on open HEVC", IEEE ICCE, pp. 65–66, Las Vegas, NV, Jan. 2014.

E163. F. Pescador et al., "Complexity analysis of an HEVC decoder based on a digital signal processor", IEEE Trans. on Consumer Electronics, vol. 59, pp. 391–399, May 2013.

E164. K. Iguchi et al., "HEVC encoder for super hi-vision", IEEE ICCE, pp. 61–62, Las Vegas, NV, Jan. 2014.

E165. K. Miyazawa et al., "Real-time hardware implementation of HEVC encoder for 1080p HD video", IEEE PCS 2013, pp. 225–228, San Jose, CA, Dec. 2013.

E166. Z. Lv and R. Wang, "An all zero blocks early detection method for high efficiency video coding" SPIE, vol. 9029, pp. 902902-1 thru 902902-7, IS&T/SPIE Electronic Imaging, San Francisco, CA, Feb. 2014. See [E171, E186].

E167. S. Vasudevan and K.R. Rao, "Combination method of fast HEVC encoding", IEEE ECTICON 2014, Korat, Thailand, May 2014.

http://www.uta.edu/faculty/krrao/dip/ click on courses and then click on EE5359 Scroll down and go to Thesis/Project Title and click on S. Vasudevan.

E168. S. Kim et al., "A novel fast and low-complexity motion estimation for UHD HEVC", IEEE PCS 2013, pp. 105–108, San Jose, CA, Dec. 2013.

E169. H. Zhang, Q. Liu and Z. Ma, "Priority classification based intra mode decision for high efficiency video coding", IEEE PCS 2013, pp. 285–288, San Jose, CA, Dec. 2013.

E170. X. Wang et al., "Paralleling variable block size motion estimation of HEVC on CPU plus GPU platform", IEEE ICME workshop, 2013.

E171. P.-T. Chiang and T.S. Chang, "Fast zero block detection and early CU termination for HEVC video coding", IEEE ISCAS 2013, pp. 1640–1643, Beijing, China, May 2013. See [E166, E186].

E172. V. Sze and M. Budagavi, "Design and implementation of next generation video coding systems", Sunday 1 June 2014 (half day tutorial), IEEE ISCAS 2014, Melbourne, Australia, 1–5 June 2014.

E173. T. Nguyen et al., 'Transform coding techniques in HEVC", IEEE Journal of selected topics in signal processing, vol. 7, pp. 978–989, Dec. 2013.

E174. E. Peixoto, T. Shanableh and E. Izquierdo, "H.264/AVC to HEVC transcoder based on dynamic thresholding and content modeling", IEEE Trans. CSVT, vol. 24, pp. 99–112, Jan. 2014.

E175. Y. Tew and K.S. Wong, "An overview of information hiding in H.264/AVC compressed video", IEEE Trans. CSVT, vol. 24, pp. 305–319, Feb. 2014.

E176. T. Na and M. Kim, "A novel no-reference PSNR estimation method with regard to deblocking filtering effect in H.264/AVC bitstreams", IEEE Trans. CSVT, vol. 24, pp. 320–330, Feb. 2014.

E177. D. Grois et al., "Performance comparison of H.265/MPEG-HEVC, VP9 and H.264/MPEG-AVC encoders", IEEE PCS 2013, pp. 394–397, San Jose, CA, Dec. 2013.

E178. F. Bossen, "Common HM test conditions and software reference configurations", document JCTVC-L1100 of JCT-VC, Geneva, CH, Jan. 2013.

E179. A. Abdelazim, W. Masri and B. Noaman, "Motion estimation optimization tools for the emerging high efficiency video coding (HEVC)", SPIE, vol. 9029, paper #1, IS&T/SPIE Electronic Imaging, pp. 902905-1 thru 902905-8, San Francisco, CA, Feb. 2014.

E180. H.R. Wu and K.R. Rao, "Digital video image quality and perceptual coding", Boca Raton, FL: CRC Press, 2006.

E181. T. Shen et al., "Ultra fast H.264/AVC to HEVC transcoder", IEEE DCC, pp. 241–250, 2013.

E182. M. Tikekar et al., "A 249-Mpixel/s HEVC video-decoder chip for 4k Ultra-HD applications", IEEE J. of Solid-State Circuits, vol. 49, pp. 61–72, Jan. 2014.

E183. K.R. Rao and J.J. Hwang, "Techniques and standards for image/video/audio coding", Prentice Hall, 1996. Translated into Japanese and Korean.

E184. B. Bross et al., "HEVC performance and complexity for 4K video", 3rd IEEE ICCE-Berlin, pp. 44–47, Sept. 2013.

E185. F. Bossen et al., "HEVC complexity and implementation analysis", IEEE Trans. CSVT, vol. 22, pp. 1685–1696, Dec. 2012.

E186. K. Lee et al., "A novel algorithm for zero block detection in high efficiency video coding", IEEE Journal of selected topics in signal processing, vol. 7, #6, pp. 1124–1134, Dec. 2013. See [E166, E171].

E187. K. Shah, "Reducing the complexity of inter prediction mode decision for high efficiency video coding", M.S. Thesis, EE Dept., University of Texas at Arlington, Arlington, Texas, May 2014. http://www.uta.edu/faculty/krrao/dip click on courses and then click on EE5359 Scroll down and go to Thesis/Project Title and click on K. Shah

E188. HEVC Fraunhofer site containing all the information on HEVC- http://hevc.info/ (software/overview papers/documents etc.)

E189. HEVC decoder for handheld devices implemented by Ace Thought- http://www.acethought.com/index.php/products/hevc-decoder/

E190. M. Revabak and T. Ebrahimi, "Comparison of compression efficiency between HEVC/H.265 and VP9 based on subjective assessments", SPIE Optical Engineering, Applications of digital image processing, Vol. 9217, San Diego, California, USA, August 18–21, 2014. https://infoscience.epfl.ch/record/200925/files/article-vp9-submited-v2.pdf

E191. D. Grois et al., "Comparative assessment of H.265/MPEG-HEVC, VP9, and H.264/MPEG-AVC encoders for low-delay video applications", SPIE Optical Engineering, Applications of digital image processing, Vol. 9217, San Diego, California, USA, August 18–21, 2014. (This has several useful web sites.)

E192. B. Bross et al., "HEVC performance and complexity for 4K video", 3rd IEEE ICCE-Berlin, Berlin, Germany. Sept. 2013.
E193. A. Heindel, E. Wige and A. Kaup, "Analysis of prediction algorithms for residual compression in a lossy to lossless scalable video coding system based on HEVC", SPIE Optical Engineering, Applications of digital image processing, Vol. 9217, San Diego, California, USA, August 18–21, 2014.
E194. A. Heindel, E. Wige and A. Kaup, "Sample-based weighted prediction for lossless enhancement layer coding in HEVC", IEEE PCS 2013, Grand Compression Challenge, 2013.
E195. J. Vanne, M. Viitanen and T.D. Hamalainen, "Efficient mode decision schemes for HEVC inter prediction", IEEE Trans. CSVT, vol. 24, pp. 1579–1593, Sept. 2014.
E196. L. Wang et al., "Lossy to lossless image compression based on reversible integer DCT", IEEE ICIP, pp. 1037–1040, San Diego, CA, Oct. 2008.
E197. K. Gruneberg, T. Schierl and S. Narasimhan, "ISO/IEC 13818-1:2013/FDAM 3, Transport of high efficiency video coding (HEVC) video over MPEG-2 systems, 2013". (FDAM: final draft amendment).
E198. F. Bossen, "Excel template for BD-rate calculation based on piecewise cubic interpolation", JCT-VC Reflector.
E199. L. Shen, Z. Zhang and Z. Liu, "Adaptive inter-mode decision for HEVC jointly utilizing inter-level and spatiotemporal correlations", IEEE Trans. CSVT, vol. 24, pp. 1709–1722, Oct. 2014.
E200. Y.-T. Peng and P.C. Cosman, "Weighted boundary matching error concealment for HEVC using block partition decisions", IEEE 48th Asilomar Conference on Signals, Systems and Computers, Pacific Grove, CA, Nov. 2014.
E201. F. Zou et al., "View synthesis prediction in the 3-D video coding extensions of AVC and HEVC", IEEE Trans. CSVT, vol. 24, pp. 1696–1708, Oct. 2014.
E202. V. Sze, M. Budagavi and G.J. Sullivan (Editors), "High efficiency video coding: Algorithms and architectures", Springer 2014.
E203. M. Warrier, "Multiplexing HEVC video with AAC audio bit streams, demultiplexing and achieving lip synchronization"' M.S. Thesis, EE Dept., University of Texas at Arlington, Arlington, Texas, Dec. 2014. http://www.uta.edu/faculty/krrao/dip click on courses and then click on EE5359 Scroll down and go to Thesis/Project Title and click on M. Warrier

E204. P. Mehta, "Complexity reduction for intra mode selection in HEVC using open MP", M.S. Thesis, EE Dept., University of Texas at Arlington, Arlington, Texas, Dec. 2014. http://www.uta.edu/faculty/krrao/dip click on courses and then click on EE5359 Scroll down and go to Thesis/Project Title and click on P. Mehta

E204a. J. Dubhashi, "Complexity reduction of H.265 motion estimation using compute unified device architecture", M.S. Thesis, EE Dept., University of Texas at Arlington, Arlington, Texas, Dec. 2014. http://www.uta.edu/faculty/krrao/dip click on courses and then click on EE5359 Scroll down and go to Thesis/Project Title and click on J. Dubhashi.

E204b. K. Suresh, "Application of open MP on complexity reduction in inter frame coding in HEVC" M.S. Thesis, EE Dept., University of Texas at Arlington, Arlington, Texas, Dec. 2014. http://www.uta.edu/faculty/krrao/dip click on courses and then click on EE5359 Scroll down and go to Thesis/Project Title and click on K. Suresh.

E205 Y. Umezaki and S. Goto, "Image segmentation approach for realizing zoomable streaming HEVC video", IEEE ICICS 2013.

E206 Y. Ye and P. Andrivon, "The scalable extensions of HEVC for ultra-high definition video delivery", IEEE Multimedia magazine, vol. 21, issue, 3, pp. 58–64, July–Sept. 2014. See the references at the end. These are highly useful for further research in SHVC – scalable HEVC.

Software Repository; Scalable Extensions of HEVC

The source code for the software is available in the following SVN repository.
https://hevc.hhi.fraunhofer.de/svn/svn_SHVCSoftware/
For tool integration branch for a company can be obtained by contacting:
Seregin, Vadim (vseregin@qti.qualcomm.com)

Build System

The software can be built under Linux using make. For Windows, solutions for different versions of Microsoft Visual Studio are provided.

Software Structure

The SHVC Test Model Software inherits the same software structure from the HEVC test model HM software, which includes the following applications and libraries for encoding, decoding and down sampling process:

- Applications:
    - TAppEncoder, executable for bit stream generation
    - TAppDecoder, executable for reconstruction.
    - TAppCommon, common functions for configuration file parsing.
    - TAppDownConvert, down sampling functionalities.
- Libraries:
    - TLibEncoder, encoding functionalities
    - TLibDecoder, decoding functionalities
    - TLibCommon, common functionalities
    - TLibVideoIO, video input/output functionalities

ATSC Advanced television systems committee www.atsc.org

TV Tomorrow: ATSC 3.0 Advances

ATSC 3.0 will be a radical departure from the current standard ATSC 1.0 was developed around 20 years ago, when cellphones were analog and streaming was unheard of. It relies 8-VSB modulation capable of delivering 19.39 Mbps in a 6 MHz TV channel – enough to carry a high – definition program compressed by a factor of 50 using MPEG-2 to a fixed receiver.

With ATSC 3.0, the committee seeks to increase that data rate by 30 percent, or roughly 25.2 Mbps. The overall intent of 3.0 is to enable seamless transmission of HD 4K, 22.2 channel audio and other data streams to fixed, mobile and handheld devices in all types of terrain.

The most differentiating characteristic of ATSC 3.0 is that it will not be backward – compatible with 1.0 or even 2.0, which is now in development. In other words, televisions now capable of processing over-the-air TV signals will not be able to decode ATSC 3.0 signals. It is also being developed with a global perspective in mind, meaning that modulations schemes other than 8-VSB – particularly co-orthogonal frequency division multiplexing, or COFDM – will likely be on the table.

In [E206], it is stated that "SHVC is a candidate technology in ATSC and it is expected to be proposed soon to DVB (digital video broadcasting) for inclusion in their codec toolbox". www.atsc.org

[E207] J. Chen et al., "Scalable HEVC (SHVC) Test Model 6 (SHM 6), JCTVC-Q1007_v1, 17th JCT-VC meeting, Valencia, ES, 27 March–4 April 2014.

[E208] L. Shen, Z. Zhang and P. An, "Fast CU size decision and mode decision algorithm for HEVC intra coding', IEEE Trans. on consumer electronics, vol. 59, pp. 207–213, Feb. 2013.

[E209] E. Kalali, Y. Adibelli and I. Hamzaoglu, "A high performance deblocking filter hardware for high efficiency video coding", IEEE Trans. on consumer electronics, vol. 59, pp. 714–720, Aug. 2013.

[E210] R. Garcia and H. Kalva, "Subjective evaluation of HEVC and AVC/H.264 in mobile environments", IEEE Trans. on Consumer Electronics, vol. 60, pp. 116–123, Feb. 2014. See also [E124].

[E211] (ATSC) Washington, D.C. 3D-TV terrestrial broadcasting, part2-SCHC using real time delivery; Doc a/104 2012, Dec 26, 2012.

[E212] S. Wang, D. Zhou and S. Goto "Motion compensation architecture for 8K HEVC decoder", IEEE ICME, Chengdu, China, 14–18 July 2014.

[E213] A. Aminlou et al., "Differential coding using enhanced inter-layer reference picture for the scalable extension of H.265/HEVC video codec", IEEE Trans. CSVT, vol. 24, pp. 1945–1856, Nov. 2014. There are several papers on SHVC in the references.

[E214] W. Hamidouche, M. Raulet and O. Deforges, "Parallel SHVC decoder: Implementation and analysis", IEEE ICME, Chengdu, China, 14–18 July 2014.

[E215] H. Wang, H. Du and J. Wu, "Predicting zero coefficients for high efficiency video coding", IEEE ICME, Chengdu, China, 14–18 July 2014.

[E216] G. Braeckman et al., "Visually lossless screen content coding using IIEVC basc-laycr", IEEE VCIP 2013, pp. 1–6, Kuching, China, 17–20, Nov. 2013.

[E217] W. Zhu et al., "Screen content coding based on HEVC framework", IEEE Trans. Multimedia, vol. 16, pp. 1316–1326 Aug. 2014 (several papers related to MRC) MRC: mixed raster coding.

[E218] ITU-T Q6/16 Visual Coding and ISO/IEC JTC1/SC29/WG11 Coding of Moving Pictures and Audio Title: Joint Call for Proposals for Coding of Screen Content Status: App. oved by ISO/IEC JTC1/SC29/WG11 and ITU-T SG16 Q6/16 (San Jose, 17 January 2014).

Final draft standard expected: late 2015.

[E219] S. Cho et al., "HEVC hardware decoder implementation for UHD video applications", IEEE ICASSP, Florence, Italy, 4–9 May 2014.

[E220] H. Kim et al., "Multi-core based HEVC hardware decoding system", IEEE ICME workshop 2014, Chengdu, China, 14–18, July 2014.

[E221] C. Yan et al., "Efficient parallel framework for HEVC motion estimation on many core processors", IEEE Trans. CSVT, vol. 24, pp. 2077–2089, Dec. 2014.

[E222] S.-F. Tsai et al., "A 1062 Mpixels/s 8192 × 4320p high efficiency video coding (H.265) encoder chip", 2013 Symposium on VLSIC, pp. 188–189, 2013.

[E223] Y. Fan, Q. Shang and X. Zeng, "In-block prediction-based mixed lossy and lossless reference frame recompression for next-generation video encoding", IEEE Trans. CSVT, vol. 25, pp. 112–124, Jan. 2015.

The following papers are presented in IEEE ICCE, Las Vegas, NV, Jan. 2015. In the ICCE proceedings, each paper is limited to 2 pages. The authors can be contacted by email for detailed papers, if any.

[E224] Real-time Video Coding System for Up to 4K 120P Videos with Spatio-temporal Format Conversion
Toshie Misu (NHK, Japan); Yasutaka Matsuo (Japan Broadcasting Corporation (NHK), Japan); Shunsuke Iwamura and Shinichi Sakaida (NHK, Japan)
A real-time UHDTV video coding system with spatio-temporal format conversion is proposed. The system pre-reduces the UHDTV source resolution to HDTV before HEVC/H.265 encoding. The decoded video is reconstructed to UHDTV in the receiver. We have developed a hardware format converter with super-resolution techniques, and have tested the coding performance. This involves down sampling (decimation), HEVC codec and upsampling (interpolation).

[E225] Rate-Distortion Optimized Composition of HEVC-Encoded Video
Weiwei Chen (Vrije Unniversiteit Brussel, Belgium); Jan Lievens and Adrian Munteanu (Vrije Universiteit Brussel, Belgium); Jürgen Slowack and Steven Delputte (Barco N. V., Belgium)
A rate control scheme is proposed to maximize the encoding quality of a video composition based on HEVC. By solving an R-D optimization problem, this scheme can optimally allocate rate for each video source and maximize the video quality of the composition. The experiments show 14.3% bitrate reduction on average.

[E226] Improvement of 8K UHDTV Picture Quality for H.265/HEVC by Global Zoom Estimation
Ryoki Takada and Shota Orihashi (Waseda University, Japan); Yasutaka Matsuo (Waseda University & Japan Broadcasting Corporation (NHK), Japan); Jiro Katto (Waseda University, Japan)
In this paper, to handle zooming in 8K video sequences, we propose a method for improving the picture quality by global zoom estimation based on motion vector analysis extracted by block matching.

[E227] Improving the Coding Performance of 3D Video Using Guided Depth Filtering
Sebastiaan Van Leuven, Glenn Van Wallendael, Robin Bailleul, Jan De Cock and Rik Van de Walle (Ghent University – iMinds, Belgium)
Autostereoscopic displays visualize 3D scenes based on encoded texture and depth information, which often lack the quality of good depth maps. This paper investigates the coding performance when applying guided depth filtering as a pre-processor. This content adaptive filtering shows gains of 5.7% and 9.3% for 3D-HEVC and MV-HEVC, respectively.

[E228] Watermarking for HEVC/H.265 Stream
Kazuto Ogawa and Go Ohtake (Japan Broadcasting Corporation, Japan)
We propose a watermarking method for HEVC/H.265 video streams that embeds information while encoding the video.

[E229] Video Copy Detection Based on HEVC Intra Coding Features
Kai-Wen Liang, Yi-Ching Chen, Zong-Yi Chen and Pao-Chi Chang (National Central University, Taiwan)
This work utilizes the coding information in HEVC for video copy detection. Both directional modes and residual coefficients of the I-frames are employed as the texture features for matching. These features are robust against different quantization parameters and different frame sizes. The accuracy is comparable with traditional pixel domain approaches.

[E230] Fast Intra Mode Decision Algorithm Based on Local Binary Patterns in High Efficiency Video Coding (HEVC)
Jong-Hyeok Lee (SunMoon University, Korea); Kyung-Soon Jang (Sunnoon University, Korea); Byung-Gyu Kim (SunMoon University, Korea); Seyoon Jeong and Jin Soo Choi (ETRI, Korea)
This paper presents a fast intra mode decision algorithm using the local binary pattern of the texture of an encoded block in HEVC

REXt. The experimental result shows 14.59% of time-saving factor on average with 0.69% of of bit-rate and 0.19 dB of PSNR loss, respectively.

[E231] Fast Thumbnail Extraction Algorithm for HEVC
Wonjin Lee (University of Hanyang, Korea); Gwanggil Jeon (University of Ottawa, Canada); Jechang Jeong (Hanyang University, Korea)
The proposed method only reconstructs 4 × 4 boundary pixels which are needed for thumbnail image. After partial decoding process, it could be concluded that the proposed method significantly reduces decoding time. In addition, the visual quality of obtained thumbnail images was almost identical to thumbnail images extracted after full decoding process.

[E232] A Probabilistic-Based CU Size Pre-Determination Method for Parallel Processing of HEVC Encoders
Taeyoung Na and Sangkwon Na (Samsung Electronics, Korea); Kiwon Yoo (Samsung. Electronics, Korea)
In this paper, CU sizes to be considered for one are pre-determined according to a probabilistic decision model instead of checking all CU size candidates before inter prediction. Thus, motion estimation (ME) with selected CU sizes can be performed in parallel and this is advantageous for H/W encoder design.

[E233] A Multicore DSP HEVC Decoder Using an Actor-Based Dataflow Model
Miguel Chavarrias, Fernando Pescador, Matias J Garrido, Eduardo Juarez and Cesar Sanz (Universidad Politécnica de Madrid, Spain)
The OpenRVC-CAL compiler framework has been used along with the OpenMP-API to implement an HEVC video decoder based on multicore DSP technology. Currently, two DSP-cores have been used, though the technique may be applied to the development of N-core based implementations. The two-core decoder outperforms a single-core implementation by 70%

[E234] A New Mode for Coding Residual in Scalable HEVC (SHVC)
Hamid Reza Tohidypour (University of British Columbia, Canada); Mahsa T Pourazad (TELUS Communications Company, Canada); Panos Nasiopoulos (University of British Columbia, Canada); Jim Slevinsky (TELUS Communications, Canada)

Recently, an effort on standardizing the scalable extension of High Efficiency Video Coding (HEVC), known as SHVC, has been initiated. To improve the compression performance of SHVC, we propose a new mode for coding residual information, which reduces the bit-rate of the enhancement layer by up to 3.13%.

[E235] Coding Efficiency of the Context-based Arithmetic Coding Engine of AVS 2.0 in the HEVC Encoder
Hyunmin Jung, Soonwoo Choi and Soo-ik Chae (Seoul National University, Korea)
This paper evaluates coding efficiency of the HEVC encoder before and after the CABAC is replaced with the CBAC while the rate errors are controlled by varying the bit depth of the fractional part for estimated bits in rate estimation tables. Experimental result shows the CBAC provides 0.2% BD-rate reduction.

[E236] Machine Learning for Arbitrary Downsizing of Pre-Encoded Video in HEVC
Luong Pham Van, Johan De Praeter, Glenn Van Wallendael, Jan De Cock and Rik Van de Walle (Ghent University – iMinds, Belgium)
This paper proposes a machine learning based transcoding scheme for arbitrarily downsizing a pre-encoded High Efficiency Video Coding video. Machine learning exploits the correlation between input and output coding information to predict the split-flag of coding units. The experimental results show that the proposed techniques significantly reduce the transcoding complexity.

[E237] Accelerating H.264/HEVC Video Slice Processing Using Application Specific Instruction Set Processor
Dipan Kumar Mandal and Mihir N. Mody (Texas Instruments, India); Mahesh Mehendale (Texas Instruments Inc., India); Naresh Yadav and Chaitanya Ghone (Texas Instruments Ltd., India); Piyali Goswami (Texas Instruments India Pvt. Ltd., India); Hetul Sanghvi (Texas Instruments Inc, India); Niraj Nandan (Texas Instruments, USA)
The paper presents a programmable approach to accelerate slice header decoding in H.264, H.265 using an Application Specific Instruction Set Processor (ASIP). Purpose built instructions accelerate slice processing by 30%–70% for typical to worst case complexity slices. This enables real time universal video decode for all slice-complexity-scenarios without sacrificing programmability.

[E238] Fast Coding Unit Size Selection for HEVC Inter Prediction
Thanuja Mallikarachchi (University of Surrey, United Kingdom); Anil Fernando (Center for Communications Research. University of Surrey, United Kingdom); Hemantha Kodikara Arachchi (University of Surrey, United Kingdom)
Determining the best partitioning structure for a given CTU is a time consuming operation within the HEVC encoder. This paper presents a fast CU size selection algorithm for HEVC using a CU classification technique. The proposed algorithm achieves an average of 67.83% encoding time efficiency improvement with negligible rate-distortion impact.

[E239] S.-J. You, S.-J. Chang and T.-S. Chang, "Fast motion estimation algorithm and design for real time QFHD high efficiency video coding", IEEE Trans. CSVT, vol. 25, pp. 1533–1544, Sept. 2015. Also many good references are listed at the end.

[E240] W. Zhao, T. Onoye and T. Song, "Hierarchical structure based fast mode decision for H.265/HEVC", IEEE Trans. CSVT, vol. 25, pp. 1651–1664, Oct. 2015. Several valuable references related to fast mode decision approaches resulting in reduced HEVC encoder complexity are cited at the end.

[E241] Y. Umezaki and S. Goto, "Image segmentation approach for realizing zoomable streaming HEVC video", ICICS 2013.

[E242] G. Correa et al., "Fast HEVC encoding decisions using data mining", IEEE Trans. CSVT, vol. 25, pp. 660–673, April 2015.

[E243] Access to HM 16.9 Software Manual:
https://hevc.hhi.fraunhofer.de/svn/svn_HEVCSoftware/tags/HM-16.9+SCM-4.0rc1/doc/software-manual.pdf

[E244] HM Encoder Description: http://mpeg.chiariglione.org/standard/mpeg-h/high-efficiency-video-coding/n14703-high-efficiency-video-coding-hevc-encoder

[E245] T. Nguyen and D. Marpe, "Objective Performance Evaluation of the HEVC Main Still Picture Profile", IEEE Trans. CSVT, vol. 25, pp. 790–797, May 2015.

[E246] C.C. Chi et al., "SIMD acceleration for HEVC decoding", IEEE Trans. CSVT, vol. 25, pp. 841–855, May 2015.

[E247] B. Min and C.C. Cheung, "A fast CU size decision algorithm for the HEVC intra encoder", IEEE Trans. CSVT, vol. 25, pp. 892–896, May 2015.

[E248] H.-S. Kim and R.-H. Park, "Fast CU partitioning algorithm for HEVC using online learning based Bayesian decision rule", IEEE Trans. CSVT, (Early access).

[E249] Y.-H. Chen and V. Sze, "A deeply pipelined CABAC decoder for HEVC supporting level 6.2 high-tier applications", IEEE Trans. CSVT, vol. 25, pp. 856–868, May 2015.

[E250] D. Zhou et al., "Ultra-high-throughput VLSI architecture of H.265/HEVC CABAC encoder for UHDTV applications", IEEE Trans. CSVT, vol. 25, pp. 497–507, March 2015.

[E251] D. Grois, B. Bross and D. Marpe, "HEVC/H.265 Video Coding Standard (Version 2) including the Range Extensions, Scalable Extensions, and Multiview Extensions," (Tutorial) Monday 29 June 2015 11.30 am–3:00 pm, IEEE ICME 2015, Torino, Italy, 29 June–3 July, 2015.

[E252] L. Hu et al., "Hardware – oriented rate distortion optimization algorithm for HEVC Intra – frame encoder", IEEE International Workshop on multimedia computing, in conjunction with IEEE ICME 2015, Torino, Italy, 29 June–3 July, 2015.

[E253] I. Hautala et al., "Programmable low-power multicore coprocessor architecture for HEVC/H.265 in-loop filtering", IEEE Trans. CSVT, vol. 25, pp. 1217–1230, July 2015.

[E254] J. KIM, S.-Ho and M. Kim, "An HEVC complaint perceptual video coding scheme based on JND models for variable block-sized transform kernels", IEEE Trans. CSVT, (early access).

[E255] F. Zou et al., "Hash Based Intra String Copy for HEVC Based Screen Content Coding, " IEEE ICME (workshop), Torino, Italy, June 2015. (In references, there are several papers on Screen Content Coding).

[E256] HEVC SCC extension reference software: https://hevc.hhi.fraunhofer.de/svn/svn HEVCSoftware/tags/HM-14.0+RExt-7.0+SCM-1.0/

[E257] C. Chen et al., "A New Block-Based Coding Method for HEVC Intra Coding," IEEE ICME (Workshop), Torino, Italy, June 2015.

[E258] L. Hu et al., "Hardware-Oriented Rate-Distortion Optimization Algorithm for HEVC Intra-Frame Encoder," IEEE ICME, Torino, Italy, June 2015.

[E259] Y. Zhang et al., "Machine Learning-Based Coding Unit Depth Decisions for Flexible Complexity Allocation in High Efficiency Video Coding," IEEE Trans. on Image Processing, Vol. 24, No. 7, pp. 2225–2238, July 2015.

[E260] D. Grois, B. Bross and D. Marpe, "HEVC/H.265 Video Coding Standard including the Range Extensions, Scalable Extensions, and Multiview Extensions," (Tutorial), IEEE ICCE, Berlin, Germany, 6–9 Sept. 2015.

[E261] S. J. Kaware and S. K. Jagtap, "A Survey: Heterogeneous Video Transcoder for H.264/AVC to HEVC", IEEE International Conference on Pervasive Computing (ICPC), pp. 1–3, Pune, India, 8–10 Jan. 2015.

[E262] Several papers related to HDR, HEVC, Beyond HEVC and SCC in "Applications of Digital Image Processing XXXVIII," HDR I session, HDR II, Applications and Extensions for High Efficiency Video Coding (HEVC), vol. 9599, SPIE. Optics + photonics, San Diego, California, USA, 9–13, Aug. 2015. Website: www.spie.org/op

[E263] E. Peixoto et al., "Fast H.264/AVC to HEVC Transcoding based on Machine Learning", IEEE International Telecommunications Symposium (ITS), pp. 1–4, 2014.

[E264] J. M. Boyce et al., "Overview of SHVC: Scalable Extensions of the High Efficiency Video Coding (HEVC) Standard," IEEE Trans. on CSVT. vol. 26, pp. 20–34, Jan. 2016.

[E265] X265: Documentations on wavefront parallel Processing. [Online] Available: http://x265.readthedocs.org/en/latest/threading.html#wavefront-parallel-processing, accessed Aug. 1, 2015

[E266] J. Kim, S.-H. Bae and M. Kim, "An HEVC – Compliant Perceptual Video Coding Scheme based on JND Models for Variable Block-sized Transform Kernels," IEEE Trans. on CSVT. (Early Access).

[E267] Y. Chen et al., "Efficient Software H.264/AVC to HEVC Transcoding on Distributed Multicore Processors," IEEE Trans. on CSVT, vol. 25, pp. 1423–1434, Aug. 2015.

[E268] H. Lee et al., "Early Skip Mode Decision for HEVC Encoder with Emphasis on Coding Quality," IEEE Trans. on Broadcasting. (Early access).

[E269] K. Lim et al., "Fast PU skip and split termination algorithm for HEVC Intra Prediction," IEEE Trans. on CSVT, vol. 25, No. 8, pp. 1335–1346, Aug. 2015.

[E270] K. Won and B. Jeon, "Complexity – Efficient Rate Estimation for Mode Decision of the HEVC Encoder," IEEE Trans. on Broadcasting. (Early Access).

[E271] L. F. R. Lucas et al., "Intra predictive Depth Map Coding using Flexible Block partitioning," IEEE Trans. on Image Processing. (Early Access).

[E272] K. Chen et al., "A Novel Wavefront – Based High parallel Solution for HEVC Encoding," IEEE Trans. on CSVT. (Early access).

[E273] J. Zhang. B. Li and H. Li, "An Efficient Fast Mode Decision Method for Inter Prediction in HEVC," IEEE Trans. on CSVT, vol. 26, PP. 1502–1515, Aug. 2016.

[E274] N. Hu and E.-H. Yang, "Fast Mode Selection for HEVC intra frame coding with entropy coding refinement based on transparent composite model," IEEE Trans. on CSVT, vol. 25, No. 9, pp. 1521–1532, Sept. 2015. ERRATUM: (early access).

[E275] S.-Y. Jou, S.-J. Chang and T.-S. Chang, "Fast Motion Estimation Algorithm and Design For Real Time QFHD High Efficiency Video Coding," IEEE Trans. on CSVT, vol. 25, pp. 1533–1544, Sept. 2015.
The papers [E276] thru [E284] were presented in IEEE ICCE Berlin, 6–9 Sept. 2015 and can be accessed from IEEEXPLORE.

[E276] N. Dhollande et al., "Fast block partitioning method in HEVC Intra coding for UHD video," IEEE ICCE, Berlin, Germany, 6–9 Sept. 2015.

[E277] W. Shi et al., "Segmental Downsampling Intra Coding Based on Spatial Locality for HEVC," IEEE ICCE, Berlin, Germany, 6–9 Sept. 2015.

[E278] S.-H. Park et al., "Temporal Correlation-based Fast Encoding Algorithm in HEVC Intra Frame Coding," IEEE ICCE, Berlin, Germany, 6–9 Sept. 2015.

[E279] R. Skupin, Y. Sanchez and T. Schirel, "Compressed Domain Processing for Stereoscopic Tile Based Panorama Streaming using MV-HEVC," IEEE ICCE, Berlin, Germany, 6–9 Sept. 2015.

[E280] A. S. Nagaraghatta et al., "Fast H.264/AVC to HEVC transcoding using Mode Merging and Mode Mapping," IEEE ICCE, Berlin, Germany, 6–9 Sept. 2015.

[E281] A. Mikityuk et al., "Compositing without Transcoding for H.265/HEVC in Cloud IPTV and VoD services," IEEE ICCE, Berlin, Germany, 6–9 Sept. 2015.

[E282] F. Chi et al., "Objective identification of most distorted coding units over UHD HEVC sequences," IEEE ICCE, Berlin, Germany, 6–9 Sept. 2015.

[E283] M. Tomida and T. Song, "Small area VLSI architecture for deblocking filter of HEVC," IEEE ICCE, Berlin, Germany, 6–9 Sept. 2015.

[E284] T. Biatek et al., "Toward Optimal Bitrate Allocation in the Scalable HEVC Extension: Application to UHDTV," IEEE ICCE, Berlin, Germany, 6–9 Sept. 2015.

[E285] References [E285] thru [E319], [E323 – E324] are presented in IEEE ICIP 2015, Quebec City, Canada

[E285] J. Mir et al., "Rate Distortion Analysis of High Dynamic Range Video Coding Techniques," IEEE ICIP, Quebec City, Canada, 27–30 Sept. 2015.

[E286] S. G. Blasi et al., "Context adaptive mode sorting for fast HEVC mode decision," IEEE ICIP, Quebec City, Canada, 27–30 Sept. 2015.

[E287] N. Hu and E.-H. Yang, "Fast inter mode decision for HEVC based on transparent composite model," IEEE ICIP, Quebec City, Canada, 27–30 Sept. 2015.

[E288] T. Laude and J. Ostermann, "Copy mode for static screen content coding with HEVC," IEEE ICIP, Quebec City, Canada, 27–30 Sept. 2015.

[E289] A. Eldeken et al., "High throughput parallel scheme for HEVC deblocking filter," IEEE ICIP, Quebec City, Canada, 27–30 Sept. 2015.

[E290] X. Shang et al.," Fast cu size decision and pu mode decision algorithm in HEVC intra coding," IEEE ICIP, Quebec City, Canada, 27–30 Sept. 2015.

[E291] Y. Chen and H. Liu, "Texture based sub-pu motion inheritance for depth coding," IEEE ICIP, Quebec City, Canada, 27–30 Sept. 2015.

[E292] Y.-C. Sun et al., "Palette mode – a new coding tool in screen content coding extensions of HEVC," IEEE ICIP, Quebec city, Canada, 27–30 Sept. 2015.

[E293] D. Grois, B. Bross and D. Marpe, "HEVC/H.265 Video Coding Standard (Version 2) including the Range Extensions, Scalable Extensions, and Multiview Extensions," (Tutorial) Sunday 27 Sept 2015, 9:00 am to 12:30 pm, IEEE ICIP, Quebec city, Canada, 27–30 Sept. 2015.

[E294] D. Schroeder et al., "Block structure reuse for multi-rate high efficiency video coding," IEEE ICIP, Quebec city, Canada, 27–30 Sept. 2015.

[E295] Y. Yan, M. M. Hannuksela and H. Li, "Seamless switching of H.265/HEVC-coded dash representations with open GOP prediction structure," IEEE ICIP, Quebec city, Canada, 27–30 Sept. 2015.

[E296] A. Sohaib and A. R. Kelly, "Radiometric calibration for HDR imaging," IEEE ICIP, Quebec city, Canada, 27–30 Sept. 2015.

[E297] J. Liu et al., "Chromatic calibration of an HDR display using 3d octree forests," IEEE ICIP, Quebec city, Canada, 27–30 Sept. 2015.

[E298] S. Patel, D. Androutsos and M. Kyan, "Adaptive exposure fusion for high dynamic range imaging," IEEE ICIP, Quebec city, Canada, 27–30 Sept. 2015.

[E299] Y. Zang, D. Agrafiotis and D. Bull, "High dynamic range content calibration for accurate acquisition and display," IEEE ICIP, Quebec city, Canada, 27–30 Sept. 2015.

[E300] J.-F. Franche and S. Coulombe, "Fast H.264 to HEVC transcoder based on post-order traversal of quadtree structure," IEEE ICIP, Quebec city, Canada, 27–30 Sept. 2015.

[E301] A. Arrufat, P. Philippe and O. Deforges, "Mode-dependent transform competition for HEVC," IEEE ICIP, Quebec city, Canada, 27–30 Sept. 2015.

[E302] W.-K. Cham and Q. Han, "High performance loop filter for HEVC," IEEE ICIP, Quebec city, Canada, 27–30 Sept. 2015.

[E303] A. Arrufat, A.-F. Perrin and P. Philippe, "Image coding with incomplete transform competition for HEVC," IEEE ICIP, Quebec city, Canada, 27–30 Sept. 2015.

[E304] Y. Liu and J. Ostermann, "Fast motion blur compensation in HEVC using fixed-length filter," IEEE ICIP, Quebec city, Canada, 27–30 Sept. 2015.

[E305] L. Yu and J. Xiao, "Statistical approach for motion estimation skipping (SAMEK)," IEEE ICIP, Quebec city, Canada, 27–30 Sept. 2015.

[E306] H.-B. Zhang et al., "Efficient depth intra mode decision by reference pixels classification in 3D-HEVC," IEEE ICIP, Quebec city, Canada, 27–30 Sept. 2015.

[E307] M. Joachimiak et al., "Upsampled-view distortion optimization for mixed resolution 3D video coding," IEEE ICIP, Quebec city, Canada, 27–30 Sept. 2015.

[E308] X. Yu et al., "VLSI friendly fast cu/pu mode decision for HEVC intra encoding: leveraging convolution neural network," IEEE ICIP, Quebec city, Canada, 27–30 Sept. 2015.

[E309] M. Hannuksela et al., "Overview of the multiview high efficiency video coding (MV-HEVC) standard," IEEE ICIP, Quebec city, Canada, 27–30 Sept. 2015.

[E310] W.-N. Lie and Y.-H. Lu, "Fast encoding of 3D color-plus-depth video based on 3D-HEVC," IEEE ICIP, Quebec city, Canada, 27–30 Sept. 2015.

[E311] M. Meddeb, M. Cagnazzo and B.-P. Popescu "ROI-based rate control using tiles for an HEVC encoded video stream over a lossy network," IEEE ICIP, Quebec city, Canada, 27–30 Sept. 2015.

[E312] W. Gueder, P. Amon and E. Steinbach, "Low-complexity block size decision for HEVC intra coding using binary image feature descriptors," IEEE ICIP, Quebec city, Canada, 27–30 Sept. 2015.

[E313] L. Du and Z. Liu, "3.975MW 18.396GBPS 2R2W SRAM for SBAC context model of HEVC," IEEE ICIP, Quebec city, Canada, 27–30 Sept. 2015.

[E314] M. Pettersson, R. Sjoberg and J. Samuelsson, "Dependent random access point pictures in HEVC," IEEE ICIP, Quebec city, Canada, 27–30 Sept. 2015.

[E315] V. Sanchez, "Lossless screen content coding in HEVC based on sample-wise median and edge prediction," IEEE ICIP, Quebec city, Canada, 27–30 Sept. 2015.

[E316] H. Maich et al., "A multi-standard interpolation hardware solution for H.264 and HEVC," IEEE ICIP, Quebec city, Canada, 27–30 Sept. 2015.

[E317] Y. Wang, F. Duanmu and Z. Ma, "Fast CU partition decision using machine learning for screen content compression," IEEE ICIP, Quebec city, Canada, 27–30 Sept. 2015.

[E318] M. Wang and K. N. Ngan, "Optimal bit allocation in HEVC for real-time video communications," IEEE ICIP, Quebec City, Canada, 27–30 Sept. 2015.

[E319] X. Lian et al., "Pixel-grain prediction and k-order ueg-rice entropy coding oriented lossless frame memory compression for motion estimation in HEVC," IEEE ICIP, Quebec City, Canada, 27–30 Sept. 2015.

[E320] Y. Shishikui, "The coming of age of 8K super hi-vision (ultra high definition television) – The meaning of the ultimate", IEEE PCS 2015, Cairns, Australia, June 2015.

[E321] D. Flynn et al., "Overview of the Range Extensions for the HEVC Standard: Tools, Profiles and Performance," IEEE Trans. CSVT, vol. 26, pp. 4–19, Jan. 2016.

[E322] J. Xu, R. Joshi and R. A. Cohen, "Overview of the Emerging HEVC Screen Content Coding Extension," IEEE Trans. CSVT, vol. 26, pp. 50–62, Jan. 2016.

[E323] Y. Yan, M. M. Hannuksela and H. Li, "Seamless switching of H.265/HEVC-coded dash representations with open GOP prediction structure," IEEE ICIP, Quebec City, Canada, 27–30 Sept. 2015.

[E324] D. Grois, B. Bross and D. Marpe, "HEVC/H.265 Video Coding Standard (Version 2) including the Range Extensions, Scalable Extensions, and Multiview Extensions," (Tutorial) Sunday 27 Sept. 2015, 9:00 am to 12:30 pm, IEEE ICIP, Quebec city, Canada, 27–30 Sept. 2015.

[E325] J.M. Boyce et al., "Overview of SHVC: Scalable extensions of the high efficiency video coding (HEVC) standard", IEEE Trans. CSVT. vol. 26, pp. 20–34, Jan. 2016.

[E326] G. Tech et al., "Overview of the Multiview and 3D extensions of high efficiency video coding", IEEE Trans. CSVT. vol. 26, pp. 35–49, Jan. 2016.

[E327] M. M. Hannuksela, J. Lainema and V. K. M. Vadakital, "The high efficiency image file format standards," IEEE Signal Processing Magazine, vol. 32, pp. 150–156, IEEE Trans. CSVT. Vol. (Early Access)

[E328] C. Li, H. Xiong and D. Wu, "Delay–Rate–Distortion Optimized Rate Control for End-to-End Video Communication Over Wireless Channels," IEEE Trans. CSVT, vol. 25, no. 10, pp. 1665–1681, Oct. 2015.

[E329] W. Hamidouche, M. Raulet and O. Deforges, "4K Real-Time and Parallel Software Video Decoder for Multi-layer HEVC Extensions," IEEE Trans. CSVT. vol. 26, pp. 169–180, Jan. 2016.

[E330] H.R. Tohidpour, M.T. Pourazad and P. Nasiopoulos, "Online Learning-based Complexity Reduction Scheme for 3D-HEVC," IEEE Trans. CSVT. (Early access).

[E331] X. Deng et al.," Subjective-driven Complexity Control Approach for HEVC," IEEE Trans. CSVT. (Early access).

[E332] S.- H. Jung and H. W. Park,"A fast mode decision method in HEVC using adaptive ordering of modes," IEEE Trans. CSVT. Vol. 26, pp. 1846–1858, Oct. 2016.

[E333] A. J. D.- Honrubia et al., "Adaptive Fast Quadtree Level Decision Algorithm for H.264 to HEVC Video Transcoding," IEEE Trans. CSVT. vol. 26, pp. 154–168, Jan. 2016.

[E334] K.-L. Chung et al., "Novel Bitrate-Saving and Fast Coding for Depth Videos in 3D-HEVC," IEEE Trans. CSVT. (Early access).

[E335] C.-H. Lin, K.-L. Chung and C.-W. Yu, "Novel Chroma Subsampling Strategy based on Mathematical Optimization for Compressing Mosaic Videos with Arbitrary RGB Color Filter Arrays in H.264/AVC and HEVC," IEEE Trans. CSVT. (Early access).

[E336] G. Correa et al., "Pareto-Based Method for High Efficiency Video Coding with Limited Encoding Time," IEEE Trans. CSVT. (Early access).

[E337] T. Vermeir et al., "Guided Chroma Reconstruction for Screen Content Coding," IEEE Trans. CSVT. (Early access).

[E338] M. Abeydeera et al., "4K Real Time HEVC Decoder on FPGA," IEEE Trans. CSVT. vol. 26, pp. 236–249, Jan. 2016.

[E339] E. Francois et al., "High Dynamic Range and Wide Color Gamut Video Coding in HEVC: Status and Potential Future Enhancements," IEEE Trans. CSVT. vol. 26, pp. 63–75, Jan. 2016.

[E340] H. Lee et al., "Fast Quantization Method with Simplified Rate-Distortion Optimized Quantization for HEVC Encoder," IEEE Trans. CSVT. vol. 26, pp. 107–116, Jan. 2016.

[E341] T.-Y. Huang and H. H. Chen, "Efficient Quantization Based on Rate-Distortion Optimization for Video Coding," IEEE Trans. CSVT. (Early access).

[E342] G. Pastuszak and A. Abramowski, "Algorithm and Architecture Design of the H.265/HEVC Intra Encoder," IEEE Trans. CSVT. vol. 26, pp. 210–222, Jan. 2016.

[E343] M. Kim et al., "Exploiting Thread-Level Parallelism on HEVC by Employing Reference Dependency Graph", IEEE Trans. CSVT. (Early access).

[E344] A. Aminlou et al., "A New R-D Optimization Criterion for Fast Mode Decision Algorithms in Video Coding and Transrating," IEEE Trans. CSVT. vol. 26, pp. 696–710, April 2016.

[E345] P.-T. Chiang et al., "A QFHD 30 fps HEVC Decoder Design," IEEE Trans. CSVT, vol. 26, pp. 724–735, April 2016.

[E346] W. Xiao et al., "HEVC Encoding Optimization Using Multicore CPUs and GPUs," IEEE Trans. CSVT. (Early access).

[E347] HEVC Demystified : A primer on the H.265 Video Codec – https://www.elementaltechnologies.com/resources/white-papers/hevc-h265-demystified-primer

[E348] N.S. Shah, "Reducing Encoder Complexity of Intra – Mode Decision Using CU Early Termination Algorithm", M.S. Thesis, EE Dept., University of Texas at Arlington, Arlington, Texas, Dec. 2015. http://www.uta.edu/faculty/krrao/dip click on courses and then click on EE5359 Scroll down and go to Thesis/Project Title and click on N.S. Shah.

[E349] D, Hingole, "H.265 (HEVC) bitstream to H.264 (MPEG 4 AVC) bitstream transcoder" M.S. Thesis, EE Dept., University of Texas at Arlington, Arlington, Texas, Dec. 2015. http://www.uta.edu/faculty/krrao/dip click on courses and then click on EE5359 Scroll down and go to Thesis/Project Title and click on D. Hingole.

[E350] U.S.M. Dayananda, "Investigation of Scalable HEVC & its bitrate allocation for UHD deployment in the context of HTTP streaming", M.S. Thesis, EE Dept., University of Texas at Arlington, Arlington, Texas, Dec. 2015. http://www.uta.edu/faculty/krrao/dip click on courses and then click on EE5359 Scroll down and go to Thesis/Project Title and click on U.S.M. Dayananda.

[E351] V. Vijayaraghavan, "Reducing the Encoding Time of Motion Estimation in HEVC using parallel programming", M.S. Thesis, EE Dept., University of Texas at Arlington, Arlington, Texas, Dec. 2015. http://www.uta.edu/faculty/krrao/dip click on courses and then click on EE5359 Scroll down and go to Thesis/Project Title and click on V. Vijayaraghavan.

[E352] S. Kodpadi, "Evaluation of coding tools for screen content in High Efficiency Video Coding", M.S. Thesis, EE Dept., University of Texas at Arlington, Arlington, Texas, Dec. 2015. http://www.uta.edu/faculty/krrao/dip click on courses and then click on EE5359 Scroll down and go to Thesis/Project Title and click on S. Kodpadi.

[E353] N.N. Mundgemane, *"Multi-stage prediction scheme for Screen Content based on HEVC",* M.S. Thesis, EE Dept., University of Texas at Arlington, Arlington, Texas, Dec. 2015. http://www.uta.edu/faculty/krrao/dip click on courses and then click on EE5359 Scroll down and go to Thesis/Project Title and click on N.N. Mundgemane.

[E354] E. de la Torre, R.R.- Sanchez and J.L. Martinez, "Fast video transcoding from HEVC to VP9", IEEE Trans. Consumer Electronics, vol. 61, pp. 336–343, Aug. 2015.

[E355] W.-S. Kim et al., "Cross-component prediction in HEVC", IEEE Trans. CSVT, [early access]

(Related papers: T. Nguyen et al., **Extended Cross-Component Prediction in HEVC**, *Proc. Picture Coding Symposium (PCS 2015)*, Cairns, Australia, May 31–Jun 03, 2015.)

A. Khairatet et al., **Adaptive Cross-Component Prediction for 4:4:4 High Efficiency Video Coding**, *Proc. IEEE International Conference on Image Processing (ICIP 2014)*, Paris, Oct. 27–30, 2014.)

Abstract in [E355] is reproduced below for ready reference.

Abstract—Video coding in the YCbCr color space has been widely used, since it is efficient for compression, but it can result in color distortion due to conversion error. Meanwhile, coding in RGB color space maintains high color fidelity, having the drawback of a substantial bit-rate increase with respect to YCbCr coding. Cross-component prediction (CCP) efficiently compresses video content by decorrelating color components while keeping high color fidelity. In this scheme, the chroma residual signal is predicted from the luma residual signal inside the coding loop.

This paper gives a description of the CCP scheme from several point-of-view, from theoretical background to practical implementation. The proposed CCP scheme has been evaluated in standardization communities and adopted into H.265/HEVC Range Extensions. Experimental results show significant coding performance improvements both for natural and screen content video, while the quality of all color components is maintained. The average coding gains for natural video are 17% and 5% bit-rate reduction in case of intra coding, and 11% and 4% in case of inter coding for RGB and YCbCr coding, respectively, while the average increment of encoding and decoding times in the HEVC reference software implementation are 10% and 4%, respectively.

[E356] C.-K. Fong, Q. Han and W.-K. Cham, "Recursive integer cosine transform for HEVC and future video coding standards", IEEE Trans. CSVT, vol. 26, (early access).

[E357] P. Hsu and C. Shen, "The VLSI architecture of a highly efficient deblocking filter for HEVC systems", IEEE Trans. CSVT, vol. 26,

Jan 2016. [early access] (Several references related to deblocking and SAO filters in HEVC and deblocking filters in H.264/AVC are listed at the end.).

[E358] H. Lee et al., "Fast quantization method with simplified rate-distortion optimized quantization for an HEVC encoder" IEEE Trans. CSVT, vol. 26, pp. 107–116, Jan. 2016.

[E359] S. Wang et al., "VLSI implementation of HEVC motion compensation with distance biased direct cache mapping for 8K UHDTV applications", IEEE Trans. CSVT, (early access).

[E360] G. Georgios, G. Lentaris and D. Reisis, "Reduced complexity super resolution for low-bitrate compression", IEEE Trans. CSVT, vol. 26, pp. 342–345, Feb. 2016. This paper can lead to several valuable and challenging projects.

[E361] C.-H. Kuo, Y.-L. Shih and S.-C. Yang, "Rate control via adjustment of Lagrangian multiplier for video coding", IEEE Trans. CSVT, pp. 2069–2078, Nov. 2016.

[E362] J. Zhang, B. Li and H. Li, "An efficient fast mode decision method for inter prediction in HEVC", IEEE Trans. CSVT, vol. 26, pp. 1502–1515, Aug. 2016.

[E363] C.-P. Fan, C.-W. Chang and S.-J. Hsu, "Cost effective hardware sharing design of fast algorithm based multiple forward and inverse transforms for H.264/AVC, MPEG-1/2/4, AVS and VC-1 video encoding and decoding applications", IEEE Trans. CSVT, (early access).

[E364] J.-S. Park et al., "2-D large inverse transform (16x16, 32X32) for HEVC (high efficiency video coding)", J. of Semiconductor Technology and Science, vol. 12, pp. 203–211, June 2012.

[E365] A. Ilic, et al., "Adaptive scheduling framework for real-time video encoding on heterogeneous systems". IEEE Trans. CSVT, (early access).

[E366] IEEE DCC, Snow Bird, Utah, 29 March–1 April 2016 Panel Presentation.

Video Coding: Recent Developments for HEVC and Future Trends

Abstract: This special event at DCC 2016 consists of a keynote talk by Gary Sullivan (co-chair of the MPEG & VCEG Joint Collaborative Team on Video Coding) followed by a panel discussion with key members of the video coding and standardization community. Highlights of the presentation include HEVC Screen Content Coding (SCC), High Dynamic Range (HDR) video coding, the Joint Exploration Model (JEM) for advances in video compression beyond HEVC, and recent initiatives in royalty-free video coding.

Panel Members:

Anne Aaron
Manager, Video Algorithms – Netflix
Arild Fuldseth
Principal Engineer, Video Coding – Cisco Systems
Marta Karczewicz
VP, Technology, Video R&D and Standards – Qualcomm
Jörn Ostermann
Professor – Leibniz Universität Hannover Institute for Information Processing
Jacob Ström
Principal Researcher – Ericsson
Gary Sullivan
Video Architect – Microsoft
Yan Ye
Senior Manager, Video Standards – InterDigital

Slides can be accessed from:

http://www.cs.brandeis.edu//~dcc/Programs/Program2016Keynoteslides-Sullivan.pdf

[E367] M. Jridi and P.K. Meher, "A scalable approximate DCT architectures for efficient HEVC compliant video coding", IEEE Trans. CSVT, (early access) See also P.K. Meher et al., "Generalized architecture for integer DCT of lengths N = 8, 16 and 32," IEEE Trans. CSVT, vol. 24, pp. 168–178, Jan. 2014.
[E368] T. Zhang et al., "Fast intra mode and CU size decision for HEVC", IEEE Trans. CSVT, (early access).
[E369] X. Liu et al., "An adaptive mode decision algorithm based on video texture characteristics for HEVC intra prediction", IEEE Trans. CSVT, (early access) Abstract is reproduced here.

Abstract—The latest High Efficiency Video Coding (HEVC) standard could achieve the highest coding efficiency compared with the existing video coding standards. To improve the coding efficiency of Intra frame, a quadtree-based variable block size coding structure which is flexible to adapt to various texture characteristics of images and up to 35 Intra prediction modes for each prediction unit (PU) is adopted in HEVC. However, the computational complexity is increased dramatically because all the possible combinations of the mode candidates are calculated in order to find the optimal rate distortion (RD) cost by using Lagrange multiplier. To alleviate the encoder computational load, this paper proposes an adaptive mode decision algorithm based on texture complexity and direction for HEVC Intra prediction. Firstly, an adaptive Coding Unit (CU) selection algorithm according to each depth levels' texture complexity is presented to filter out unnecessary coding blocks. And then, the original redundant mode candidates for each PU are reduced according to its texture direction. The simulation results show that the proposed algorithm could reduce around 56% encoding time in average while maintaining the encoding performance efficiently with only 1.0% increase in BD-rate compared to the test mode HM16 of HEVC.

There are also several references related to intra coding both in HEVC and H.264/AVC.

[E370] Z. Pan et al., "Early termination for TZSearch in HEVC motion estimation", IEEE ICASSP 2013, pp. 1389–1392, June 2013.

[E371] R. Shivananda, "Early termination for TZSearch in HEVC motion estimation", project report, Spring 2016. http://www.uta.edu/faculty/krrao/dip click on courses and then click on EE5359 Scroll down and go to projects Spring 2016 and click on R. Shivananda final report. See [E370]. Following this technique, Shivananda is able to reduce HEVC encoding time by 38% with negligible loss in RD performance for LD, RA and RA early profiles. See Tables 3 thru 14 for the results.

[E372] S.K. Rao, "Performance comparison of HEVC intra, JPEG, JPEG 2000, JPEG XR, JPEG LS and VP9 intra", project report, Spring 2016. http://www.uta.edu/faculty/krrao/dip click on courses and then click on EE5359 Scroll down and go to projects Spring 2016 and click on Swaroop Krishna Rao final report.

[E373] H.N. Jagadeesh, "Sample Adaptive Offset in the HEVC standard", project report, Spring 2016. http://www.uta.edu/faculty/krrao/dip click on courses and then click on EE5359 Scroll down and go to projects Spring 2016 and click on Jagadeesh final report. He has implemented HM software using HEVC test sequences with and without the SAO filter. In general SAO filter improves the subjective and objective qualities of the video output at the cost of slight increase in implementation complexity.

[E374] N. Thakur, "Fast Intra Coding Based on Reference Samples Similarity in HEVC", project report Spring 2016. http://www.uta.edu/faculty/krrao/dip click on courses and then click on EE5359 Scroll down and go to projects Spring 2016 and click on Nikita Thakur final report.

[E375] M.N. Sheelvant, "Performance and Computational Complexity Assessment of High-Efficiency Video Encoders", project report Spring 2016. http://www.uta.edu/faculty/krrao/dip click on courses and then click on EE5359 Scroll down and go to projects Spring 2016 and click on M.N. Sheelvant final report.

[E376] Y. Tew, K.S. Wong and R.C.-W. Phan, "Region of interest encryption in high efficiency video coding compressed video", IEEE- ICCE Taiwan, May 2016.

[E377] S.-H. Bae, J. Kim and M. Kim, "HEVC-Based perceptually adaptive video coding using a DCT-Based local distortion detection probability", IEEE Trans. on Image Processing, vol. 25, pp. 3343–3357, July 2016.

[E378] C. Diniz et al., "A deblocking filter hardware architecture for the high efficiency video coding standard", Design, Automation and Test in Europe conference and exhibition (DATE), 2015.

[E379] D.S. Pagala, "Multiplex/demultiplex of HEVC video with HE-AAC v2 audio and achieve lip synch.", M.S. Thesis, EE Dept., University of Texas at Arlington, Arlington, Texas, Aug. 2016. http://www.uta.edu/faculty/krrao/dip click on courses and then click on EE5359 Scroll down and go to Thesis/Project Title and click on D.S. Pagala.

[E380] Sessions related to HEVC in IEEE ICIP, Phoenix, Arizona, Sept. 2016:

MA-L3: SS: Image compression grand challenge,
Several sessions on Image quality assessment
MPA.P3: Image and Video Coding.
MPA.P8: Transform Coding.
TA-L1: HEVC Intra Coding.
TP-L1: HEVC optimization
WPB-P2 HEVC processing and coding

Panel Session: Is compression dead or are we wrong again?

[E381] M. Masera, M. Martina and G. Masera, "Adaptive approximated DCT architectures for HEVC", IEEE Trans. CSVT, (early access). This has several references related to integer DCT architectures.

[E382] B. Min, Z. Xu and C.C. Cheung, "A fully pipelined hardware architecture for intra prediction of HEVC", IEEE Trans. CSVT (early access). The fully pipelined design for intra prediction in HEVC can produce 4 pels per clock cycle. Hence the throughput of the proposed architecture is capable of supporting 3840 × 2160 videos at 30 fps.

[E383] G. Correa et al., "Pareto-based method for high efficiency video coding with limited encoding time", IEEE Trans. CSVT, vol. 26, pp. 1734–1745, Sept. 2016.

[E384] C.-H. Liu, K.-L. Chung and C.-YW. Yu, "Novel Chroma Subsampling Strategy Based on Mathematical Optimization for Compressing Mosaic Videos With Arbitrary RGB Color Filter Arrays in H.264/AVC and HEVC", IEEE Trans. CSVT, vol. 26, pp. 1722–1733, Sept. 2016.

[E385] C. Herglotz et al., "Modeling the energy consumption of HEVC decoding process", IEEE Trans. CSVT, (early access).

[E386] IEEE Data Compression Conference (DCC) Snowbird, Utah, April 4–7, 2017. http://www.cs.brandeis.edu/~dcc

Keynote address
"Video Quality Metrics", S. Daly, Senior Member Technical Staff, Dolby Laboratories
Special sessions:
"Video coding", G. Sullivan and Y. Ye, Session Chairs
"Genome Compression", J. Wen and K. Sayood, Session Chairs
(note there is Session 10 on Genome Compression)
"Quality Metrics and Perceptual Compression" Y. Reznik and T. Richter, Session Co-Chairs
Web site below.
http://www.cs.brandeis.edu/~dcc/Programs/Program2016KeynoteSlides-Sullivan.pdf
There are also several papers related to HEVC.

[E387] H. Siddiqui, K. Atanassov and S. Goma, "Hardware-friendly universal demosaick using non-iterative MAP reconstruction", IEEE ICIP 2016 (MA.P.3.4), Phoenix, AZ, Sept. 2016.

[E388] J. Seok et al., "Fast Prediction Mode Decision in HEVC Using a Pseudo Rate-Distortion Based on Separated Encoding Structure", ETRI J. vol. 38, #5, pp. 807–817, Oct. 2016.

Abstract is reproduced below.
A novel fast algorithm is suggested for a coding unit (CU) mode decision using pseudo rate-distortion based on a separated encoding structure in High Efficiency Video Coding (HEVC). A conventional HEVC encoder requires a large computational time for a CU mode prediction because prediction and transformation procedures are applied to obtain a rate-distortion cost. Hence, for the practical application of HEVC encoding, it is necessary to significantly reduce the computational time of CU mode prediction. As described in this paper, under the proposed separated encoder structure, it is possible to decide the CU prediction mode without a full processing of the prediction and transformation to obtain a rate-distortion cost based on a suitable condition. Furthermore, to construct a suitable condition to improve the encoding speed, we employ a pseudo rate-distortion estimation based on a Hadamard transformation and a simple quantization. The experimental results show that the proposed method achieves a 38.68% reduction in the total encoding time with a similar coding performance to that of HEVC reference model.

This is an excellent paper which describes the complexity involved in implementing the HM software and develops techniques that reduce the encoder complexity significantly with minimal loss in PSNR and BD rate supported by simulation results (4K and 2K test sequences).

[E389] T. Mallikarachchi et al., "Content-Adaptive Feature-Based CU Size Prediction for Fast Low-Delay Video Encoding in HEVC", IEEE Trans. CSVT, (early access).

[E390] J. Zhou et al., "A Variable-Clock-Cycle-Path VLSI Design of Binary Arithmetic Decoder for H.265/HEVC" IEEE Trans. CSVT, (early access).

[E391] A. Zheng et al., "Adaptive Block Coding Order for Intra Prediction in HEVC", IEEE Trans. CSVT, vol. 26, pp. 2152–2158, Nov. 2016.

[E392] G. Fracastoro, S.M. Fosson and E. Magli, "Steerable discrete cosine transform", IEEE Trans. IP, vol. 26, pp. 303–314, Jan. 2017. This has several references related to DDT, MDDT and graph based transforms. See also G. Fracastoro and E. Magli, "Steerable discrete cosine transform," in Proc. IEEE International Workshop on Multimedia Signal Processing, 2015 (MMSP), 2015.

[E393] Q. Huang et al., "Understanding and Removal of False Contour in HEVC Compressed Images", IEEE Trans. CSVT, (Early access). Abstract is repeated here.

Abstract—A contour-like artifact called false contour is often observed in large smooth areas of decoded images and video. Without loss of generality, we focus on detection and removal of false contours resulting from the state-of-the-art HEVC codec. First, we identify the cause of false contours by explaining the human perceptual experiences on them with specific experiments. Next, we propose a precise pixel-based false contour detection method based on the evolution of a false contour candidate (FCC) map. The number of points in the FCC map becomes fewer by imposing more constraints step by step. Special attention is paid to separating false contours from real contours such as edges and textures in the video source. Then, a decontour method is designed to remove false contours in the exact contour position while preserving edge/texture details. Extensive experimental results are provided to demonstrate the superior performance of the proposed false contour detection and removal (FCDR) method in both compressed images and videos.

[E394] F. Chen et al., "Block-composed background reference for high efficiency video coding", IEEE Trans. CSVT, (Early access).

[E395] D. Schroeder et al., "Efficient multi-rate video encoding for HEVC-based adaptive HTTP streaming," *IEEE Trans. CSVT, Early Access.* Part of abstract is repeated here:

> "We propose methods to use the encoding information to constrain the rate-distortion-optimization of the dependent encodings, so that the encoding complexity is reduced, while the RD performance is kept high. We additionally show that the proposed methods can be combined, leading to an efficient multi-rate encoder that exhibits high RD performance and substantial complexity reduction. Results show that the encoding time for 12 representations at different spatial resolutions and signal qualities can be reduced on average by 38%, while the average bitrate increases by less than 1%"

[E396] L. Li et al., "Domain optimal bit allocation algorithm for HEVC", IEEE Trans. CSVT, (Early access).

[E397] B. Min, Z. Xu and R.C. Cheung, "A fully pipelined hardware architecture for intra prediction of HEVC", IEEE Trans. CSVT, (Early access).

[E398] P.-K. Hsu and C.-A. Shen, "The VLSI architecture of a highly efficient deblocking filter for HEVC system", IEEE Trans. CSVT, (Early access). There are many other references related to deblocking filter.

[E399] R. Wang et al., "MPEG internet video coding standard and its performance evaluation", IEEE Trans. CSVT, (Early access).

This standard approved as ISO/IEC (MPEG-4 Internet Video Coding) in June 2015 is highly beneficial to video services over the internet. This paper describes the main coding tools used in IVC and evaluates the objective and subjective performances compared to web video coding (WVC), video coding for browsers (VCB) and AVC high profile.

[E400] M. Xu et al., "Learning to detect video saliency with HEVC features", IEEE Trans. IP, vol. 26, pp. 369–385, Jan. 2017. Introduction is repeated here.

According to the study on the human visual system (HVS) [1], when a person looks at a scene, he/she may pay much visual attention on a small region (the fovea) around a point of eye fixation at high resolutions. The other regions, namely the peripheral regions, are captured with little attention at low resolutions. As such, humans are able to avoid is therefore a key to perceive the world around humans, and it has been extensively studied in psychophysics, neurophysiology, and even computer vision sciences [2]. Saliency detection is an effective way to predict the amount of human visual attention attracted by different regions in images/videos, with computation on their features. Most recently, saliency detection has been widely applied in object detection [3, 4], object recognition [5], image retargeting [6], image quality assessment [7], and image/video compression [8, 9].

[E401] T. Mallikarachchi et al., "A feature based complexity model for decoder complexity optimized HEVC video encoding", IEEE ICCE, Las Vegas, Jan. 2017.

[E402] F. Pescador et al., "Real-time HEVC decoding with openHEVC and openMP", IEEE ICCE, Las Vegas, Jan. 2017.

[E403] G. Xiang et al., "An improved adaptive quantization method based on perceptual CU early splitting for HEVC", IEEE ICCE, Las Vegas, Jan. 2017.

[E404] H.M. Maung, S. Aramvith and Y. Miyanaga, "Error resilience aware rate control and mode selection for HEVC video transmission", IEEE ICCE, Las Vegas, Jan. 2017.
[E405] B. Boyadjis et al., "Extended selective encryption of H.264/AVC (CABAC) and HEVC encoded video streams", IEEE Trans. on CSVT, vol. 27, pp. 892–906, April 2017. Conclusion is repeated here.

Conclusion

This paper proposes a novel method for SE of H.264/AVC (CABAC) and HEVC compressed streams. Our approach tackles the main security challenge of SE: the limitation of the information leakage through protected video streams.

The encryption of luma prediction modes for Intra blocks/units is proposed in addition to the traditional ES in order to significantly improve the *structural* deterioration of video contents. Especially on I frames, it brings a solid complement to previous SE schemes relying on residual encryption (mainly affecting the *texture* of objects) and motion vector encryption (disrupting the *movement* of objects). The proposed scheme shows both numerical and visual improvements of the scrambling performance regarding state-of-the-art SE schemes. Based on the encryption of *regular*-mode treated syntax element, the approach exhibits a loss in compression efficiency that has been estimated empirically up to 2% on I frames, which remains, in a lot of application cases, a totally tolerable variation.

In this paper, we complemented the state of the art analysis proposed in [15] of the impact of each cipherable syntax element in HEVC standard with the proposed Intra prediction encryption. The study confirms the premise that targeting this specific element is particularly efficient from the SE perspective: the PSNR and SSIM variations we obtain tend to demonstrate that it is the most impacting code word in the proposed ES. Considering their role in the video reconstruction, such a result is not really surprising: their encryption heavily impacts Intra frames but also Intra blocks/units in Inter frames, and the generated error naturally drifts both spatially and temporally within the video stream.

We did not provide in this paper an exhaustive analysis of CABAC regular mode cipherable syntax elements. Some of them are easily identifiable (for example, the process proposed in this paper could almost identically apply to the Intra prediction modes for Chroma components), but a large majority of these cipherable elements would require a very specific monitoring which

we did not address in our study. An analysis of the *maximum* ES for both H.264/AVC and HEVC, based on the proposed *regular* mode encryption, is thus one major track for future research. Nevertheless, such a study would have to pay a particular attention to a critical trade- off we highlighted in our study; the underlying correlation between the improvement of the scrambling performance and its consequences on the compression efficiency.
See http://www.ieee.org/publications_standards/publications/rights/index.html for more information.

[E406] JVET documents are archived at http://phenix.int-evry.fr/jvet coding tools are defined to some extent.intent may change prior to final publication. Citation information: DOI 10.1109/TCSVT.2015.2511879, (restricted access).

[E407] J. An et al., “Block partitioning structure for next generation video coding”, ITU-T SG16, Doc. COM16-C966, Sept. 2015.

[E408] H. Huang et al., “EE2.1: Quadtree plus binary structure integration in JEM tools”, JVET-C0024, Geneva, May 2016.

[E409] S.-H. Park and E.S. Jang, “An efficient motion estimation method for QTBT structure in JVET”, IEEE DCC, April 2017.

[E410] P.-L. Cabarat, W. Hamidouche and O. Deforges, “Real-time and parallel SHVC hybrid codec AVC to HEVC decoder”, IEEE ICASSP 2017, pp. 3046–3050, New Orleans, March 2017.

[E411] K. Senzaki et al., “BD-PSNR/Rate Computation tool for five data points”, document JCTVC-B055, Geneva, Switzerland, July 2010.

[E412] Y. Fan et al., “A hardware oriented IME algorithm for HEVC and its hardware implementation”, IEEE Trans. CSCT, Early access.

[E413] T. Zhang et al., “Signal dependent transform based on SVD for HEVC intra coding”, IEEE Trans. Multimedia (early access). Section VI conclusions is repeated here.

This paper investigated the residuals characteristics of intra prediction and the deficiency of DCT/DST for residual blocks. Based on the residual analysis, a signal dependent transform based on SVD (SDT-SVD) which is derived from the synthetic block is proposed to compact the residuals from angular intra prediction. The proposed SDT-SVD is also extended to template matching prediction. Experimental results suggest that the proposed coding method with SDT-SVD for AIP outperforms the latest HEVC reference software with a bit rate reduction of 1.0% on average and it can be up to 2.1%.

When SDT-SVD is extended to TMP based intra coding, the overall bit rate reduction is 2.7% on average and can be up to 5.8%. In the future, fast algorithms, such as early skip for SDT-SVD transform and fast template searching method, can be exploited to reduce the complexity of the proposed method. To further improve the efficiency of the SDT-SVD transform, more work can be done to design an efficient block with more similar structures as the residual block for deriving SVD transforms.

E414 X. Lu et al., "Fast intra coding implementation for high efficiency video coding", IEEE DCC 2017.

IEEE ICCE, Las Vegas, NV, 8–11 Jan. 2016 (ICCE international conference on consumer electronics)

Papers on HEVC: (These papers can lead to several projects.)

ICCE1. Error Resilience Aware Motion Estimation and Mode Selection for HEVC Video Transmission

Gosala Kulupana, Dumidu S. Talagala, Hemantha Kodikara Arachchi and Anil Fernando (University of Surrey, United Kingdom)

Error concealment techniques such as motion copying require significant changes to HEVC (High Efficiency Video Coding) motion estimation process when incorporated in error resilience frameworks. This paper demonstrates a novel motion estimation mechanism incorporating the concealment impact from future coding frames to achieve an average 0.73dB gain over the state-of-the-art.

ICCE 2. H.721: Standard for Multimedia Terminal Devices Supporting HEVC

Fernando M Matsubara (Mitsubishi Electric Corporation, Japan); Hideki Yamamoto (Oki, Japan)

The newest ITU-T H.721 standard "IPTV terminal devices: Basic model" is described. Main goal of this revision is to enhance content quality by adopting latest video coding and adaptive streaming techniques. Selected technologies include HEVC and MPEG-DASH. This paper summarizes the standard and carefully chosen profiles to maximize perceived quality.

ICCE3. A Downhill Simplex Approach for HEVC Error Concealment in Wireless IP Networks.

Kyoungho Choi (Mokpo National University, Korea); Do Hyun Kim (ETRI, Korea)

In this paper, a novel video error concealment algorithm is presented for HEVC in wireless IP networks. In the proposed approach, a downhill simplex approach is adopted for fine-tuning motion vectors, considering residual errors and block reliability, and minimizing boundary errors along prediction blocks at the same time.

ICCE4. Power Consumption Comparison Between H.265/HEVC and H.264/AVC for Smartphone Based on Wi-Fi, 3G and 4G Networks.

Min Xu (California State University, Long Beach, USA); Xiaojian Cong (California State University Long Beach, USA); Qinhua Zhao (California State University Long Beach, USA); Hen-Geul Yeh (California State University Long Beach, USA)

The H.265/HEVC can help reduce much bandwidth for streaming video on mobile networks where wireless spectrum is at a premium. However, it will take more computing power for decoding. This paper is to perform a power consumption evaluation of streaming and decoding H.265/HEVC and H.264/AVC video for smartphones.

ICCE5. A New Scene Change Detection Method of Compressed and Decompressed Domain for UHD Video Systems

Yumi Eom (Seoul National University of Science and Technology, Korea); Sang-Il Park (Korea Communications Commission, Korea); Chung Chang Woo (Seoul National University of Science and Technology, Korea)

We propose a new method using two layers that can detect the scene change frames in a fast and accurate way. Also, we propose an algorithm that can be applied to the HEVC. Through the experimental results, our proposed method will help UHD video contents analysis and indexing.

ICCE6. Improvement of H.265/HEVC Encoding for 8K UHDTV by Detecting Motion Complexity

Shota Orihashi (Waseda University & Graduate School of Fundamental Science and Engineering, Japan); Harada Rintaro (Waseda University & Fundamental Science and Engineering, Japan); Yasutaka Matsuo (Japan Broadcasting Corporation (NHK), Japan); Jiro Katto (Waseda University, Japan)

We propose a method to improve H.265/HEVC encoding performance for 8K UHDTV moving pictures by detecting amount or complexity of object motions. The proposed method estimates motion complexity by external process, and selects an optimal prediction mode and search ranges of motion vectors for highly efficient and low computation encoding.

ICCE7. Design of Multicore HEVC Decoders Using Actor-based Dataflow Models and OpenMP

Miguel Chavarrias, Matias J Garrido and Fernando Pescador (Universidad Politécnica de Madrid, Spain); Maxime Pelcat (INSA Rennes, France); Eduardo Juarez (Universidad Politcnica de Madrid, Spain)

This paper explains a backend for the RVC compiler framework. This backend uses OpenMP instead of the pthreads to automatically generate code for multicore architectures. Implementations of an HEVC decoder have been automatically generated for three multicore architectures. The backend doesn't introduce a performance penalty regarding to the C backend.

ICCE8. Perceptual Distortion Measurement in the Coding Unit Mode Selection for 3D-HEVC

Sima Valizadeh (University of British Columbia, Canada); Panos Nasiopoulos (The University of British Columbia, Canada); Rabab Ward (University of British Columbia, Canada)

In this paper, we propose to integrate a perceptual video quality metric inside the rate distortion optimization process of the 3D-HEVC. Specifically, in the coding unit (CU) mode selection process, PSNR-HVS is used as a measure for distortion. Our proposed approach improves the compression efficiency of the 3D-HEVC.

ICCE9. CTU Level Decoder Energy Consumption Modelling for Decoder Energy-Aware HEVC Encoding

Thanuja Mallikarachchi (University of Surrey & University of Surrey, United Kingdom); Hemantha Kodikara Arachchi, Dumidu S. Talagala and Anil Fernando (University of Surrey, United Kingdom)

Accurate modelling of the decoding energy of a CTU is essential to determine the appropriate level of quantization required for decoder energy-aware video encoding. The proposed method predicts the number of nonzero DCT coefficients, and their energy requirements with an average accuracy of 4.8% and 11.19%, respectively.

ICCE10. Real-time Encoding/Decoding of HEVC Using CUDA Capable GPUs

Tony James (Smartplay Technologies Pvt Ltd, Georgia)

HEVC is latest video standard from JCT-VC. It offers better compression performance compared to H.264. However, the computational complexity is highly demanding for real-time applications especially at UHD resolutions. In this paper, we propose few optimization methods to achieve real-time encoding/decoding on CUDA capable GPUs. Experimental results validate our proposal.

ICCE11. SATD Based Intra Mode Decision Algorithm in HEVC/H.265

Jongho Kim (ETRI, Korea)

HEVC which is the next generation video coding standard provides up to 35 intra prediction modes to improve the coding efficiency. These various prediction modes bring a high computation burden. In this paper, the fast intra mode decision algorithm is proposed.

ICCE12. Content Dependent Intra Mode Selection for Medical Image Compression Using HEVC

Saurin Parikh (Florida Atlantic University & Nirma University, USA); Damian Ruiz (Universitat Politécnica de Valéncia, Spain); Hari Kalva (Florida Atlantic University, USA); Gerardo Fernandez (University of Castilla La Mancha, Spain)

This paper presents a method for complexity reduction in medical image encoding that exploits the structure of medical images. The HEVC lossless intra coding of medical images of CR modality, shows reduction upto 52.47%

in encoding time with a negligible penalty of 0.22%, increase in compressed file size.

ICCE13. Decoding of Main 4:2:2 10-bit Bitstreams in HEVC Main 8-bit Best-Effort Decoders

Woo-Seok Jeong, Hyunmin Jung and Soo-ik Chae (Seoul National University, Korea)

This paper describes a new method of best-effort decoding that substantially reduces PSNR loss especially in intra prediction. For intra prediction in the decoder, reference pixels are decoded in the 4:2:2 10-bit format and non-reference pixels are in the 4:2:0 8-bit format.

http://x265.org

This is the HEVC encoder developed by Multicoreware Inc. The main difference from HM is that x265 is fast compared to HM, they have assembly level optimizations up to avx2 so far.

https://x265.readthedocs.org/en/default/api.html

This is the link to their documentation. Even ffmpeg has x265 integrated in itself, but I don't think we can use x265 in ffmpeg, since it will take a while to find where x265 is implemented in ffmpeg and then modify it and build it.

The following papers are presented in SPIE PHOTONICS WEST, February 2015, San Francisco, CA

Link: http://spie.org/EI/conferencedetails/real-time-image-video-processing

1. **Efficient fast thumbnail extraction algorithm for HEVC**

   Paper 9400–15

   Author(s): Wonjin Lee, Hanyang Univ. (Korea, Republic of); Gwanggil Jeon, Univ. of Incheon (Korea, Republic of); Jechang Jeong, Hanyang Univ. (Korea, Republic of)

The proposed algorithm is fast thumbnail extraction algorithm for HEVC. The proposed method only reconstructs 4 $\times$ 4 boundary pixels which are needed for thumbnail image. The proposed method generates the thumbnail image without full decoding. The proposed algorithm can reduce the computational complexity by performing only calculation needed for thumbnail image extraction. The proposed algorithm can significantly reduce the computational complexity for intra prediction and inverse transform. Also, it does not perform de-blocking filter and sample adaptive offset (SAO). After partial decoding process, it could be concluded that the proposed method significantly reduces the decoding time. In addition, the visual quality of obtained thumbnail images was almost identical to thumbnail images extracted after full decoding process.

2. **A simulator tool set for evaluating HEVC/SHVC streaming**

   Paper 9400–22

   Author(s): James M. Nightingale, Tawfik A. Al Hadhrami, Qi Wang, Christos Grecos, Univ. of the West of Scotland (United Kingdom); Nasser Kehtarnavaz, The Univ. of Texas at Dallas (United States)

Since the delivery of version one of H.265, the Joint Collaborative Team on Video Coding have been working towards standardization of a scalable extension (SHVC), a series of range extensions and new profiles. As these enhancements are added to the standard the range of potential applications and research opportunities will expand. For example the use of video is also growing rapidly in other sectors such as safety, security, defense and health with real-time high quality video transmission playing an important role in areas like critical infrastructure monitoring and disaster management, each of which may benefit from the application of enhanced HEVC/H.265 and SHVC capabilities. The majority of existing research into HEVC/H.265 transmission has focused on the consumer domain with the lack of freely available tools widely cited as an obstacle to conducting this type of research. In this paper we present a toolset that facilitates the transmission and evaluation of HEVC/H.265 and SHVC encoded video on an open source emulator. Our toolset provides researchers with a modular, easy to use platform for evaluating video transmission and adaptation proposals on large-scale wired, wireless and hybrid architectures. The proposed toolset would significantly facilitate further research in delivering adaptive video streaming based on the latest video coding standard over wireless mesh and other ad hoc networks.

3. **Subjective evaluation of H.265/HEVC based dynamic adaptive video streaming over HTTP (HEVC-DASH)**

   Paper 9400–24

   Author(s): Iheanyi C. Irondi, Qi Wang, Christos Grecos, Univ. of the West of Scotland (United Kingdom)

With the surge in Internet video traffic, real-time HTTP streaming of video has become increasingly popular. Especially, applications based on the MPEG Dynamic Adaptive Streaming over HTTP standard (DASH) are emerging for adaptive Internet steaming in response to the unstable network conditions. Integration of DASH streaming technique with the new video coding standard H.265/HEVC is a promising area of research in light of the new codec's promise of substantially reducing the bandwidth requirement. The performance of such HEVC-DASH systems has been recently evaluated using objective metrics such as PSNR by the authors and a few other researchers. Such objective evaluation is mainly focused on the spatial fidelity of the pictures whilst the impact of temporal impairments incurred by the nature of reliable TCP communications is also noted. Meanwhile, subjective evaluation of the video quality in the HEVC-DASH streaming system is able to capture the perceived video quality of end users, and is a new area when compared with the counterpart subjective studies for current streaming systems based on H.264-DASH. Such subjective evaluation results will shed more light on the Quality of Experience (QoE) of users and overall performance of the system. Moreover, any correlation between the QoE results and objective performance metrics will help designers in optimizing system performance. This paper presents a subjective evaluation of the QoE of a HEVC-DASH system implemented in a hardware testbed. Previous studies in this area have focused on using the current H.264/AVC or SVC codecs and have hardly considered the effect of Wide Area Network (WAN) characteristics. Moreover, there is no established standard test procedure for the subjective evaluation of DASH adaptive streaming. In this paper, we define a test plan for HEVC demonstrate the bitrate switching operations in response to various network condition patterns. The testbed consists of real-world servers (web server and HEVC-DASH server), a WAN emulator and a real-world HEVC-DASH client. -DASH with a carefully justified data set taking into account longer video sequences that would be sufficient to evaluate the QoE by investigating the perceived impact of various network conditions such as different packet loss rates and fluctuating bandwidth, and the perceived quality of using different DASH video stream segment sizes on a video streaming session and using different video content types. Furthermore, we demonstrate the temporal structure and impairments as identified by previous objective quality metrics and capture how they are perceived by the subjects. The Mean Opinion Score (MOS) is employed and a beyond MOS evaluation method is designed based on a questionnaire that gives more insight into the performance of the system and the expectation of the users. Finally, we explore the correlation between the MOS and the objective metrics and hence establish optimal HEVC-DASH operating conditions for different video streaming scenarios under various network conditions.

NAB2014, Las Vegas, NV April 2014 (following 3 papers) (National Association of Broadcasters)

NAB is held every year in Las Vegas, NV generally in April. It is a B2B convention with nearly 10,000 attendees and has innumerable products related to broadcasting.

1. J. Pallett, (Telestream, Inc.) "HEVC; comparing MPEG-2, H.264 and HEVC".
2. S. Kumar, (Interra systems) "Aspects of video quality assessment for HEVC compressed format".
3. Y. Ye (Interdigital), "HEVC video coding".

NAB2015, Las Vegas, NV, April 2015 (following are the papers related to HEVC) (National Association of Broadcasters)

1. The New Phase of Terrestrial UHD Services: Live 4K UHD Broadcasting via Terrestrial Channel.
2. The Combination of UHD, MPEG-DASH, HEVC and HTML5 for New User Experiences.
3. Implications of High Dynamic Range on the Broadcast Chain for HD and Ultra-HD Content.
4. Beyond the Full HD (UHD & 8K).

NTT-electronics is developing HEVC/H.265 based ASICs and codec systems scheduled for mid 2015.
Digital video and systems business group.
http://www.ntt-electronics.com/en

Special Issues on HEVC

Spl H1. Special issue on emerging research and standards on next generation video coding, IEEE Trans. CSVT, vol. 22, pp.1646–1909, Dec. 2012.

Spl H2. Introduction to the issue on video coding: HEVC and beyond, IEEE Journal of selected topics in signal processing, vol. 7, pp. 931–1151, Dec. 2013.

Spl H3. Call for papers, IEEE transactions on circuits and systems for video technology. Special Section on Efficient HEVC Implementation

(M. Budagavi, J.R. Ohm, G.J. Sullivan, V. Sze and T. Wiegand Guest editors).

Spl H4. Special Issue on New Developments in HEVC Standardization and Implementations Approved by the CSVT editorial board, July 2014.

Spl H5. IEEE Journal on Emerging and Selected Topics in Circuits and Systems (JETCAS) Special Issue on Screen Content Video Coding and Applications. Vol. 6, issue 4, pp. 389–584 Dec. 2016

Spl H6. Journal of real-time image processing: Special issue on architectures and applications of high efficiency video coding (HEVC) standard for real time video applications. B.-G. Kim, K. Psannis and D.-S. Jun (Editors).

Spl H7. Special issue on Broadcasting and Telecommunications Media technology, ETRI journal, Deadline: 12/31/2015 and Publication: Oct. 2016. Website: http://etrij.etri.re.kr

Spl H8. Special issue on HEVC extensions and efficient HEVC implementations, IEEE Trans. CSVT, vol. 26, pp.1–249, Jan. 2016.

HEVC Overview Online

S. Riabstev, "Detailed overview of HEVC/H.265", [online]. Available: https://app.box.com/s/ rxxxzr1a1lnh7709yvih (accessed on June 12 2014).

HEVC tutorial by I.E.G. Richardson: http://www.vcodex.com/h265.html

JCT-VC Documents can be found in JCT-VC document management system

http://phenix.int-evry.fr/jct (see [E185])

All JCT-VC documents can be accessed. [Online]. Available: http://phenix.int-evry.fr/jct/doc_end_user/current_meeting.php?id_meeting=154&type_order=&sql_type=document_number

VCEG & JCT documents available from

http://wftp3.itu.int/av-arch in the video-site and jvt-site folders (see [E185])

HEVC encoded bit streams

ftp://ftp.kw.bbc.co.uk/hevc/hm-11.0-anchors/bitstreams/

Test Sequences

TS.1 http://media.xiph.org/video/derf/
TS.2 http://trace.eas.asu.edu/yuv/
TS.3 http://media.xiph.org/

TS.4 http://www.cipr.rpi.edu/resource/sequences/

TS.5 HTTP://BASAKOZTAS.NET

TS.6 www.elementaltechnologies.com – 4K Video Sequences
4K (3840 × 2160) UHD video test sequences
Multimedia communications with SVC and HEVC
http://r2d2n3po.istory.com/50

TS.7 Elemental 4K Test Clips.: [online] Available: http://www.elementaltechnologies.com/resources/4k-test-sequences, accessed Aug. 1, 2014

TS.8 Harmonic 4K Test Clips: [online] Available: http://www.harmonicinc.com/resources/videos/4k-video-clip-center, accessed Aug. 1, 2014.

TS.9 Kodak Lossless True Color Image Suite. [online]. Available: http://r0k.us/graphics/kodak/

TS.10 Test sequences from dropbox
https://www.dropbox.com/sh/87snmt6j70qquyb/AADuHPbf1o9Y-YnB8N4lCt-Na?dl=0

TS.11 F. Bossen, "Common test conditions and software reference configuration", JCT-VC-1100, July 2012 (Access to test sequences and QP values).

TS.12 Test Sequences: ftp://ftp.kw.bbc.co.uk/hevc/hm-11.0-anchors/bitstreams/

TS.13 Link to Screen content Test Sequences: http://pan.baidu.com/share/link?shareid=3128894651&uk=889443731

TS.14 Images of camera captured content: http://www.lifeisspiritual.com/wp-content/uploads/2014/06/Peace-Nature-Jogging.jpg

TS.15 Image of a video clip with text overlay http://www.newscaststudio.com/graphics/bloomberg/

TS.16 Video test sequences – http://forum.doom9.org/archive/index.php/t-135034.html

TS.17 https://hevc.aes.tu-berlin.de/web/testsuite-HEVC test sequences
European Broadcasting Union "EBU UHD-1 Test Set", 2013 Available online

1. To access these sequences free there is an order form (to be filled) subject to the rules and regulations. https://tech.ebu.ch/testsequences/uhd-1

Thanks for the hint, it seems that EBU change the location, please check the new location that I found on the EBU website: https://tech.ebu.ch/Jahia/site/tech/cache/offonce/testsequences/uhd-1
BR, Benjamin

On Sep 14, 2014, at 10:54 PM, "Rao, Kamisetty R" <rao@uta.edu> wrote:
In one of your papers you cited : EBU UHD-1 Test Set (Online) http://tech.edu.ch/testsequences/uhd-1
I am unable to access this web site. Please help me on this. I like to access this set for my research group. Thanks.

TS.18 http://mcl.usc.edu/
The MCL-JCI dataset consists of 50 source images with resolution 1920 × 1080 and 100 JPEG-coded images for each source image with the quality factor (QF) ranging from 1 to 100.
Here is the link to some useful videos on HEVC standard
http://www.youtube.com/watch?v=YQs9CZY2MXM
You might also find this useful and may subscribe to the same
http://vcodex.com/h265.html
HEVC/H.265 PRESENTATION (Krit Panusopone)
www.box.com/s/rxxxzr1a1lnh7709yvih

TS19 The New Test Images – Image Compression Benchmark," *Rawzor-Lossless Compression Software for Camera RAW Images [Online]*, Available: http://imagecompression.info/test images/.
Implementation of Intra Prediction in the HM reference encoder: https://hevc.hhi.fraunhofer.de/svn/svn_HEVCSoftware/tags/HM-12.0/
HM 10.1 SOFTWARE
https://hevc.hhi.fraunhofer.de/svn/svn_HEVCSoftware/branches/HM-10.1-dev/
Latest is HM16.9
Software reference manual for HM. https://hevc.hhi.fraunhofer.de/svn/svn_HEVCSoftware/branches/HM-16.9-dev/doc/software-manual.pdf

TS20 "The new test images – Image compression benchmark", http://imagecompression.info/test_images
TS21 BOSSbase image data set 512 × 512 test images
IEEE Jo*BossBase-1.01-hugo-alpha=0.4.tar.bz2*. [Online]. Available: http://agents.fel.cvut.cz/stegodata/ accessed Mar. 2014.

TS22 http://www3.americanradiology.com/pls/web1/wwimggal.vmg

TS23 K. Seshadrinathan, A. C. Bovik, and L. K. Cormack. (2010). *LIVE Video Quality Database*. [Online].
Available: http://live.ece.utexas.edu/research/quality/live_video.html

TS24 S. Péchard, R. Pépion, and P. Le Callet. *IRCCyN/IVC Videos 1080i Database*. [Online]. Available: http://www.irccyn.ec-nantes.fr/~lecallet/platforms.htm, accessed Oct. 16, 2015.

TS25 F. Zhang, S. Li, L. Ma, Y. C. Wong, and K. N. Ngan. IVP Subjective Quality Video Database. [Online]. Available: http://ivp.ee.cuhk.edu.hk/research/databas/subjective/ accessed Oct. 16, 2015.

TS26 J.-S. Lee *et al.*. (2010). *MMSP Scalable Video Database*. [Online]. Available: http://mmspg.epfl.ch/svd

TS27 J. Guo et al., "Response to B1002 Call for Test Materials: Five Test Sequences for Screen Content Video Coding May 2016." Document JVET-C0044, ITU-T SG 16 WP3 and ISO/IEC JTC1/SC29/WG 11, May 2016.

TS28 https://testimages.org

The TESTIMAGES archive is a huge and free collection of sample images designed for analysis and quality assessment of different kinds of displays (i.e. monitors, televisions and digital cinema projectors) and image processing techniques. The archive includes more than 2 million images originally acquired and divided in three different categories: SAMPLING and SAMPLING_PATTERNS (*aimed at testing resampling algorithms*), COLOR (*aimed at testing color rendering on different displays*) and PATTERNS (*aimed at testing the rendering of standard geometrical patterns*).

IEEE Journal on Emerging and Selected Topics in Circuits and Systems (JETCAS)

Special Issue on Screen Content Video Coding and Applications

Screen content video has evolved from a niche to a mainstream due to the rapid advances in mobile and cloud technologies. Real-time, low-latency transport of screen visuals between devices in the form of screen content video is becoming prevalent in many applications, e.g. wireless display, screen mirroring, mobile or external display interfacing, screen/desktop virtualization and cloud gaming. Today's commonly-used video coding methods, however, have been developed primarily with camera-captured content in mind. These new applications create an urgent need for efficient coding of screen content video, especially as the support of 4k or even 8k resolution begins to achieve mass market appeal.

Screen content video coding poses numerous challenges. Such content usually features a mix of computer generated graphics, text, and camera-captured images/video. With their distinct signal characteristics, content adaptive coding becomes necessary. This is without mentioning the varied level of the human's visual sensitivity to distortion in different types of content; visually or mathematically lossless quality may be required for all or part of the video.

Recognizing the demand for an industry standard for coding of screen content, the ISO/IEC Moving Picture Experts Group and ITU-T Video Coding Experts Group have since January 2014 been developing new extensions for HEVC. The Video Electronics Standards Association also recently completed a Display Stream Compression (DSC) standard for next-generation mobile or TV/Computer display interfaces. The development of these standards introduced many new ideas, which are expected to inspire more future innovations and benefit the varied usage of screen content coding.

Besides coding, there are many other challenging aspects related to screen content video. For instance, in applications like screen mirroring and screen/desktop virtualization, low-latency video processing and transmission are essential to ensure an immediate screen response. In addition to real-time streaming technologies, these applications need a parallel-friendly screen encoding algorithm that can be performed efficiently on modern mobile devices or remote servers in the data center, and require, in certain use cases, the harmony of their computing resources, to keep the processing time to a minimum. At the receiver side, best-effort decoding with consideration for transmission errors, along with visual quality enhancement, is expected. Addressing these constraints requires research from multiple disciplines as is the case for other applications.

The intent of this special issue is to present the latest developments in standards, algorithms, and system implementations related to the coding and processing of screen content video. Original and unpublished research results with topics in any of the following areas or beyond are hereby solicited.

- Screen content video coding techniques and standards, e.g. HEVC extensions and DSC
- Visually or mathematically lossless screen content video coding
- Application-specific screen content coding, e.g. display stream or frame memory compression
- Screen-content related pre/post-processing, e.g. resizing and post-filtering
- Visual quality assessment for screen content video

- Parallel-friendly, low-delay encoding optimization
- Robust decoding with error and power control
- Hardware/software/cloud-based screen codec implementations
- Real-time, adaptive screen content transport over Internet or wireless networks
- Design examples of novel screen content video applications, e.g. screen/desktop virtualization and cloud gaming
- System performance analysis and characterization

[SCC15] IEEE Journal on Emerging and Selected Topics in Circuits and Systems (JETCAS)

Special Issue on Screen Content Video Coding and Applications, vol. 6, issue 4, pp. 389–584, Dec. 2016.

See the overview paper: W.-H. Peng, et al., "Overview of screen content video coding: Technologies, standards and beyond", IEEE JETCAS, vol. 6, issue 4, pp. 393–408, Dec. 2016.

S.-H. Tsang, Y.-L. Chan and W.-C. Siu, "Fast and Efficient Intra Coding Techniques for Smooth Regions in Screen Content Coding Based on Boundary Prediction Samples", ICASSP2015, Brisbane, Australia, April. 2015.

J. Nam, D. Sim and I.V. Bajic, "HEVC-based Adaptive Quantization for Screen Content Videos," IEEE Int. Symposium on Broadband Multimedia Systems, pp. 1–4, Seoul, Korea, 2012.

How to Access JCT-VC Documents

JCT-VC DOCUMENTS can be found in JCT-VC document management system http://phenix.int-evry.fr/jct

All JCT-VC documents can be accessed. [Online].

http://phenix.int-evry.fr/jct/doc_end_user/current_meeting.php?id_meeting=154&type_order=&sql_type=document_number

Accessing JCT-VC Document Database

- In an internet browser, visit http://phenix.int-evry.fr/jct/.
- Create an account by clicking the link shown in the web page.

- Enter your details, and put THE UNIVERSITY OF TEXAS AT ARLINGTON in the "organization" field and also put in the appropriate motivation like STUDIES/COURSEWWORK, etc.,
- Check for email suggesting the registration was successful.
- You are good to access the content like ALL MEETINGS and NEXT MEETING.

VCEG documents http://wftp3.itu.int video site folder)

HM-16.9 Software – > https://hevc.hhi.fraunhofer.de/svn/svn_HEVCSoftware/tags/HM-16.9/

HM-16.9 Software Manual – > https://hevc.hhi.fraunhofer.de/svn/svn_HEVCSoftware/tags/HM-16.9/doc/software-manual.pdf

ElecardHEVCAnalyser: http://www.elecard.com/en/products/professional/analysis/hevc-analyzer.html

Scalable Extension of HEVC – > https://hevc.hhi.fraunhofer.de/svn/svn_HEVCSoftware

Encoding time evaluation: Intel VTune Ampllfier XE Software profiler, Available http://software.intel.com (accessed May 6, 2014).

See reference 31 in [E242]

Subjective Evaluation of Compression Algorithms and Standards

SE.1 P. Hanhart and T. Ebrahimi, "Calculation of average coding efficiency based on subjective quality scores", J. VCIR, vol. 25, pp. 555–564, April 2014. This is a very interesting and valuable paper on subjective quality and testing. The references listed at the end are highly useful. A MATLAB implementation of the proposed model can be downloaded from http://mmspg.epfl.ch/scenic This paper can lead to several projects (EE5359 Multimedia Processing).

SE.2 H.R. Wu et al., "Perceptual visual signal compression and transmission", Proc. IEEE, vol. 101, pp. 2025–2043, Sept. 2013.

SE.3 C. Deng et al. (Editors), "Visual Signal Quality Assessment: Quality of Experience (QoE)", Springer, 2015.

SE.4 T.K. Tan et al., "Video Quality Evaluation Methodology and Verification Testing of HEVC Compression Performance", IEEE Trans.

CSVT, vol. 26, pp. 76–90, Jan. 2016. Abstract of this paper is reproduced here for ready reference.

Abstract—The High Efficiency Video Coding (HEVC) standard (ITU-T H.265 and ISO/IEC 23008-2) has been developed with the main goal of providing significantly improved video compression compared with its predecessors. In order to evaluate this goal, verification tests were conducted by the Joint Collaborative Team on Video Coding of ITU-T SG 16 WP 3 and ISO/IEC JTC 1/SC 29. This paper presents the subjective and objective results of a verification test in which the performance of the new standard is compared with its highly successful predecessor, the Advanced Video Coding (AVC) video compression standard (ITU-T H.264 and ISO/IEC 14496-10). The test used video sequences with resolutions ranging from 480p up to ultra-high definition, encoded at various quality levels using the HEVC Main profile and the AVC High profile. In order to provide a clear evaluation, this paper also discusses various aspects for the analysis of the test results. The tests showed that bit rate savings of 59% on average can be achieved by HEVC for the same perceived video quality, which is higher than a bit rate saving of 44% demonstrated with the PSNR objective quality metric. However, it has been shown that the bit rates required to achieve good quality of compressed content, as well as the bit rate savings relative to AVC, are highly dependent on the characteristics of the tested content.

This paper has many valuable references including subjective quality assessment methods recommended by ITU-T.

SE5 J.-S. Lee and T. Ebrahimi, "Perceptual video compression: A survey," IEEE J. Selected Topics on Signal Process., vol. 6, no. 6, pp. 684–697, Oct. 2012.

SE6 P. Hanhart et al., "Subjective quality evaluation of the upcoming HEVC video compression standard", SPIE Applications of digital image processing XXXV, vol. 8499, paper 8499–30, Aug. 2012.

SE7 W. Lim and J.C. Kuo, "Perceptual video quality metrics: a survey", J. VCIR, vol. 22, pp. 297–312, 2011.

SE8 F. Zhang and D.R. Bull, "A Perception-based Hybrid Model for Video Quality Assessment", IEEE Trans. CSVT, vol. 26, pp. 1017–1028, June 2016.

SE9 Y. Li et al., "No-reference image quality assessment using statistical characterization in the shearlet domain." Signal Processing: Image Communication, vol. 29, pp. 748–759, July 2014.

SE10 Y. Li et al., “No reference image quality assessment with shearlet transform and deep neural networks,” Neurocomputing, vol. 154, pp. 94–109, 2015.

SE11 Y. Li et al., “No Reference Video Quality Assessment with 3D shearlet Transform and Convolutional Neural Networks,” IEEE Trans. on CSVT, vol. 26, pp. 1044–1047, June 2016.

SE12 Y. Li, et al., “No-reference image quality assessment using shearlet transform and stacked auto encoders,” IEEE ISCAS, pp. 1594–1597, May 2015.

SE13 T. Daede, “Video codec testing and quality measurement,” https://tools.ietf.org/html/draft-daede-netvc-testing, 2015.

SE14 ITU-R BT2022 (2012) General viewing conditions for subjective assessment of quality of SDTV and HDTV television pictures on at panel displays. International Telecommunication Union.

SE15 ITU-R BT500-13 (2012) Methodology for the subjective assessment of the quality of television pictures. International Telecommunication Union.

SE16 Q. Wu et al., “Blind image quality assessment using local consistency aware retriever and uncertainty aware evaluator”, IEEE Trans. CSVT, Early access.

SE17 W. Lin and C.C. Jay-Kuo, “Perceptual visual quality metrics: A survey”, J. Visual Communication Image Representation, vol. 22, pp. 297–312, May 2011.

SE18 C.-C. Jay Kuo, “Perceptual coding: Hype or Hope?”, 8th International Conference on QoMEX, Keynote, Lisbon, Portugal, 6–8 June 2016.

SE19 S.-H. Bae and M. Kim, “DCT-QM; A DCT-Based Quality Degradation Metric for Image Quality Optimization Problems ”, IEEE Trans. IP, vol. 25, pp. 4916–4930, Oct. 2016. This has several references related to image quality assessment.

SE20 IEEE Data Compression Conference (DCC), Snowbird, Utah, April 4–7, 2017. http://www.cs.brandeis.edu/∼dcc, Keynote address “Video Quality Metrics” by Scott Daly, Senior Member Technical Staff, Dolby Laboratories.

SE21 F.M. Moss et al., “On the Optimal Presentation Duration for Subjective Video Quality Assessment”, IEEE Trans. CSVT, vol. 26, pp. 1977–1987, Nov. 2016.

Apart from the comprehensive list of references related to subjective VQA, based on the DSCQS tests on video sequences the authors [SE21] conclude

"There is a small but significant increase in accuracy if sequences are increased from 1.5 to 5, 7, or 10 s.
2) This effect becomes stronger if the difference in distortion between the reference and test video is reduced.
3) The main effect remains consistent between different but temporally consistent source videos.
4) Observers feel just as confident assessing the quality of videos that are 5 s as ones that are 10 s.
The practical implications of these findings are significant.
Our results indicate that critical observations of video quality do not significantly change if 10-s sequences are exchanged for 7-, or indeed, 5-s sequences." Limitations on these findings are elaborated in the conclusions.

SE22 S. Hu et al., "Objective video quality assessment based on perceptually weighted mean square error", IEEE Trans. CSVT, (early access).

SE23 S. Wang et al., "Subjective and objective quality assessment of compressed screen content images", IEEE JETCAS, vol. 6, issue 4, pp. 532–543, Dec. 2016.

SE24 H. Yang Y. Fang W. Lin Z. Wang "Subjective quality assessment of screen content images" Proc. Int. Workshop Quality Multimedia Experience (QoMEX), pp. 257–262 Sept. 2014.

SE25 S. Shi et al, "Study on subjective quality assessment of screen content images" IEEE Proc. Picture Coding Symp. (PCS), pp. 75–79 June 2015.

SE26 New strategy for image and video quality assessment

J. Electron. Imaging, Vol. 19, 011019 (2010); doi:10.1117/1.3302129
Published 18 February 2010
ABSTRACT
REFERENCES (27)

Qi Ma, Liming Zhang, and Bin Wang
Fudan University, Department of Electronic Engineering, No. 220 Handan Road, Shanghai, China

Image and video quality assessment (QA) is a critical issue in image and video processing applications. General full-reference (FR) QA criteria such as peak signal-to-noise ratio (PSNR) and mean squared error (MSE) do not accord well with human subjective assessment. Some QA indices that consider human visual sensitivity, such as mean structural similarity (MSSIM) with structural sensitivity, visual information fidelity (VIF) with statistical sensitivity, etc., were proposed in view of the differences between reference and distortion frames on a pixel or local level. However, they ignore the role of human visual attention (HVA). Recently, some new strategies with HVA have been proposed, but the methods extracting the visual attention are too complex for real-time realization. We take advantage of the phase spectrum of quaternion Fourier transform (PQFT), a very fast algorithm we previously proposed, to extract saliency maps of color images or videos. Then we propose saliency-based methods for both image QA (IQA) and video QA (VQA) by adding weights related to saliency features to these original IQA or VQA criteria. Experimental results show that our saliency-based strategy can approach more closely to human subjective assessment compared with these original IQA or VQA methods and does not take more time because of the fast PQFT algorithm.

SE27 l. Xu et al., "Multi-task learning for image quality assessment", IEEE Trans. CSVT (early access).

SE28 K. Ma, "Waterloo exploration database" New challenges for image quality assessment models", IEEE Trans. IP, (early access).
This paper has extensive list of references related to IQA and also on image data bases. The authors suggest four different directions in terms of future work.

SE29 **The following papers are presented in IEEE DCC 2017, April, 2017.**

SESSION 7, *Quality Metrics and Perceptual Compression, Part 1*

9:40am: Reduced Reference Image Quality Assessment Based on Entropy of Classified Primitives 231

Zhaolin Wan1, Yutao Liu1, Feng Qi2, and Debin Zhao1

1Harbin Inst. of Tech, 2Inst. of Computing Tech.

10:00 am: Revisiting Perceptual Distortion for Natural Images: Mean Discrete Structural Similarity Index 241

Christopher Hillar1 and Sarah Marzen2

1Univ. of California, Berkeley, 2 Massachusetts Inst. of Tech.

10:20 am: Semantic Perceptual Image Compression Using Deep Convolution Networks. 250

Aaditya Prakash, Nick Moran, Solomon Garber, Antonella Dilillo, and James Storer

Brandeis University

SE30 R. Fang, R.A. – Bayaty and D. Wu, "BNB method for no refer ence image quality assessment", IEEE Trans. CSVT, vol. 27, pp. 1381–1391, July 2017.
(BNB: blurriness, noisiness and blockiness)

SE31 Q. Huynh-Thu and M. Ghanbari, "Scope of validity of PSNR in image/video quality assessment", IET Electronics Letters, vol. 44, (13): 800, 2008.

SE-P1 Bae and Kim have developed the DCT-Q Based Quality Degradation Metric for Image Quality Optimization Problem that has both high consistency with perceived quality and mathematically desirable properties. See [SE19] for details. The authors conclude that the DCT-QM can effectively be used not only for objective IQA tasks with unknown distortion types but also for image quality optimization problems with known distortion types. They also state "We plan to apply our DCT-QM for image/video coding and processing applications for the improvement of coding efficiency and perceptual visual quality." Explore this. Note that DCT-QM serves for still images only.

SE-P2 See SE-P1. Note that the performance comparison of HEVC (DCT-QM based HM16.0 and conventional HM16.0 for low delay configuration) is shown in Figure 9 for test sequences BasketballDrill, BQ Mall, PartyScene and RaceHorse). See also Figures 10 and 11. Please extend this comparison to HM 16.0 random access configuration.

SE-P3 See [SE19] The authors state "Full application of DCT-QM for specific codecs require extending DCT-QM for specific transform block sizes, which will be part of our future work". They imply other than 4 × 4-sized DCT kernel i.e., 8 × 8, 16 × 16, 32 × 32 and even 64 × 64. Consider specifically how the different sized DCT kernels influence the DCT-QM. Consider also the implementation complexity.

Books on HEVC

Book1. V. Sze, M. Budagavi and G.J. Sullivan (Editors), "High efficiency video coding: Algorithms and architectures", Springer 2014.

Book2. M. Wien, "High Efficiency Video Coding: Coding Tools and Specification", Springer, 2014.

Book3. I.E. Richardson, "Coding video: A practical guide to HEVC and beyond", Wiley, 2017.

Book4. K.R. Rao, D.N. Kim and J.J. Hwang, "Video coding standards: AVS China, H.264/MPEG-4 Part10, HEVC, VP6, DIRAC and VC-1", Springer, 2014.
Translated into Spanish by Dr. Carlos Pantsios M, Professor Titular/Telecommunicaciones, USB/UCA/UCV, Dept. of Electronica & Circuits, Simon Bolivar University, Caracas, Venejuela. Also published in Chinese by China Machine press – approved by Springer.

Book5. S. Wan and F. Yang, "New efficient video coding – H.265/HEVC – Principle, standard and application", in Chinese, Publishing house of electronic industry, http://www.phei.com.cn, 2014.

Book6. S. Okubo (Editor-in-Chief), "H.265/HEVC Textbook", in Japanese, Impress, Japan, 2013.

Book7. W. Gao and S. Ma, "Advanced Video Coding Systems," Springer, 2015.

Book8. M. Kavanagh, "H.265/HEVC Overview and Comparison with H.264/AVC," Sold by: Amazon Digital Services, Inc. August, 2015.

Book9. B. Bing, "Next-Generation Video Coding and Streaming," Wiley, Hoboken, NJ, 2015.

Book10. J.-R. Ohm, "Multimedia signal coding and transmission," Springer, 2015.

Book11. A.N. Netravali and B.G. Haskell, "Digital pictures, representation and compression", Plenum Press, 1988. II Edition, Springer 1995.

Book12. D. Grois et al., "High Efficiency Video Coding: A Guide to the H.265/HEVC standard," Cambridge University Press, UK, 2017.

Book13. M. E. Al-Mualla, C. N. Canagarajah and D. R. Bull, "Video Coding for Mobile Communications: Efficiency, Complexity and Resilience," Academic Press, Orlando, FL, 2002.

Book14. Y.Q. Shi and H. Sun, "Image and video compression for multimedia engineering: Fundamentals, algorithms and standards", CRC Press, 1999.

Book15. J.-B. Lee and H. Kalva, "The VC-1 and H.264 video compression standards for broadband video services", Springer 2008.

Overview Papers

OP1. G.J. Sullivan and J.-R. Ohm, "Recent developments in standardization of high efficiency video coding (HEVC)," Proc. SPIE, Applications of Digital Image Processing XXXIII, vol. 7798, pp. 77980V-1 through V-7, San Diego, CA, 1–3 Aug. 2010.

OP2. G.J. Sullivan et al., "Overview of the high efficiency video coding (HEVC) standard", IEEE Trans. CSVT, vol. 22, pp. 1649–1668, Dec. 2012.

OP3. HM9 High efficiency video coding (HEVC) Test Model 9 (HM 9) encoder description JCTVC-K1002v2, Shanghai meeting, Oct. 2012.

OP4. M.T. Pourazad et al., "HEVC: The new gold standard for video compression", IEEE Consumer Electronics Magazine, vol. 1, issue 3, pp. 36–46, July 2012.

OP5. G.J. Sullivan et al., "Standardized extensions of high efficiency video coding (HEVC)", IEEE Journal of selected topics in signal processing, vol. 7, pp. 1001–1016, Dec. 2013.

OP6. J. M. Boyce et al., "Overview of SHVC: Scalable Extensions of the High Efficiency Video Coding (HEVC) Standard," IEEE Trans. on CSVT. vol. 26, pp. 20–34, Jan. 2016.

OP7. D. Flynn et al., "Overview of the Range Extensions for the HEVC Standard: Tools, Profiles and Performance," IEEE Trans. CSVT. vol. 26, pp. 4–19, Jan. 2016.

OP8. J. Xu, R. Joshi and R. A. Cohen, "Overview of the Emerging HEVC Screen Content Coding Extension," IEEE Trans. CSVT. vol. 26, pp. 50–62, Jan. 2016.

OP9. G. Tech et al., "Overview of the Multiview and 3D extensions of high efficiency video coding", IEEE Trans. CSVT. vol. 26, pp. 35–49, Jan. 2016.

OP10. Detailed overview of HEVC/H.265: https://app.box.com/s/rxxxzr1a1lnh7709yvih

OP11. W.-H. Peng et al., "Overview of screen content video coding: Technologies, Standards and Beyond", IEEE JETCAS, vol. 6, issue 4, pp. 393–408, Dec. 2016. Abstract is reproduced below as a review.

Abstract

This paper presents recent advances in screen content video coding, with an emphasis on two state-of-the-art standards: HEVC/H.265 Screen Content Coding Extensions (HEVC-SCC) by ISO/IEC Moving Picture Experts Group and ITU-T Video Coding Experts Group, and Display Stream Compression (DSC) by Video Electronics Standards Association. The HEVC-SCC enhances the capabilities of HEVC in coding screen content, while DSC provides lightweight compression for display links. Although targeting different application domains, they share some design principles and are expected to become the leading formats in the marketplace in the coming years. This paper provides a brief account of their background, key elements, performance, and complexity characteristics, according to their final specifications. As we survey these standards, we also summarize prior arts in the last decade and explore future research opportunities and standards developments in order to give a comprehensive overview of this field.

The list of references [OP11] at the end is highly comprehensive covering not only SCC, but also related areas such as BTC, JPEG, JPEG XS, MCC, MRC, CIC, LZ, SE, VQ, SSIM among others.

Tutorials

Tut1. N. Ling, "High efficiency video coding and its 3D extension: A research perspective," Keynote Speech, ICIEA, Singapore, July 2012.
Tut2. X. Wang et al., "Paralleling variable block size motion estimation of HEVC on CPU plus GPU platform", IEEE ICME workshop, 2013.
Tut3. H.R. Tohidpour, M.T. Pourazad and P. Nasiopoulos, "Content adaptive complexity reduction scheme for quality/fidelity scalable HEVC", IEEE ICASSP 2013, pp. 1744–1748, June 2013.
Tut4. M. Wien, "HEVC – coding tools and specifications", Tutorial, IEEE ICME, San Jose, CA, July 2013.
Tut5. D. Grois, B. Bross and D. Marpe, "HEVC/H.265 Video Coding Standard (Version 2) including the Range Extensions, Scalable Extensions, and Multiview Extensions," (Tutorial) Monday 29 June 2015 11.30 am–3:00 pm, IEEE ICME 2015, Torino, Italy, 29 June–3 July, 2015.
Tut6. D. Grois, B. Bross and D. Marpe, "HEVC/H.265 Video Coding Standard including the Range Extensions, Scalable Extensions, and Multiview Extensions," (Tutorial), IEEE ICCE, Berlin, Germany, 6–9 Sept. 2015.
Tut7. D. Grois, B. Bross and D. Marpe, "HEVC/H.265 Video Coding Standard (Version 2) including the Range Extensions, Scalable Extensions, and Multiview Extensions," (Tutorial) Sunday 27 Sept 2015, 9:00 am to 12:30 pm), IEEE ICIP, Quebec City, Canada, 27–30 Sept. 2015.

This tutorial is for personal use only.
Password: a2FazmgNK
https://datacloud.hhi.fraunhofer.de/owncloud/public.php?service=files&t=8edc97d26d46d4458a9c1a17964bf881

Tut8. Please find the links to YouTube videos on the tutorial – HEVC/H.265 Video Coding Standard including the Range Extensions, Scalable Extensions and Multiview Extensions below:
https://www.youtube.com/watch?v=TLNkK5C1KN8

Tut9. HEVC tutorial by I.E.G. Richardson, www.vcodex.com/h265.html

Tut10. C. Diniz, E. Costa and S. Bampi, "High Efficient Video Coding (HEVC): From Applications to Low-power Hardware Design", IEEE ISCAS, Montreal, Canada, 22-25 May 2016. Abstract is reproduced below:

Digital video applications are widespread in every consumer electronic devices. Ultra-high resolutions videos (UHD, e.g. 8k × 4k and 4k × 2k resolutions) are gaining importance in the market. The deve- lopment of an improved video coding standard with higher efficiency for higher resolution led to the High Efficiency Video Coding (HEVC), H.265, published in 2013. This new video coding standard reaches up to approximately double compression efficiency of H.264/AVC standard for similar video quality, due to its sophisticated block partitioning schemes and novel coding algorithms. Its higher compression efficiency comes with a significant increase in computational effort in the video codecs. Real-time HEVC encoding of UHD resolution videos is a challenge, especially considering mobile video capable hardware devices that must consume lower energy to increase battery life. This energy efficiency requirement for future multimedia processors is requiring hardware architecture innovations to integrate multi-core processors with many on-chip hardware accelerators for compute-intensive tasks of the video encoder/decoder. This tutorial covers the algorithms and the dedicated hardware accelerators which are more energy efficient than general purpose processors in performing video tasks. Hardware accelerators will be shown in the tutorial, either as dedicated (ASIC) architectures or as configured/reconfigured FPGA designs. The tutorial starts with an overview the basic concepts on video representation and video coding, before moving to the details of the new HEVC algorithms, data structures, and features. We present a detailed analysis of HEVC video coding application to identify the most compute-intensive tasks of video codec. The second part of the tutorial covers the algorithmic optimization for dedicated hardware design and implementation. The state-of-the-art hardware architectures for video codec blocks are presented. In the end we point to significant future challenges to design low-power HEVC video codec systems.

Keywords: Video Coding, HEVC, Compression Algorithms, Hardware Architecture Design.

Tut11. Introduction to Motion estimation and Motion compensation—> http://www.cmlab.csie.ntu.edu.tw/cml/dsp/training/coding/motion/me1.html
HEVC presentation:
http://www.hardware.fr/news/12901/hevc-passe-ratifie.html
Detailed overview of HEVC/H.265:
https://app.box.com/s/rxxxzr1a1lnh7709yvih
HEVC white paper-Ittiam Systems:
http://www.ittiam.com/Downloads/en/documentation.aspx
HEVC white paper-Elemental Technologies:
http://www.elementaltechnologies.com/lp/hevc-h265-demystified-white-paper
HEVC white paper – http://www.ateme.com/an-introduction-to-uhdtv-and-hevc

HEVC/H.265 Video Coding Standard including the Range Extensions Scalable Extensions and..

Share your videos with friends, family, and the world

Watch now...

https://www.youtube.com/watch?v=V6a1AW5xyAw

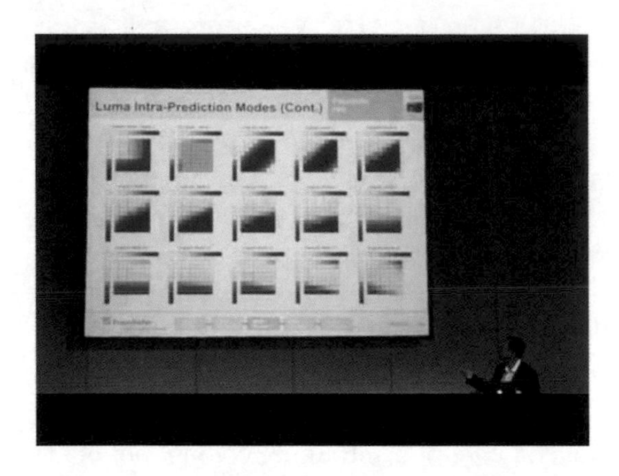

HEVC/H.265 Video Coding Standard including the Range Extensions Scalable Extensions and..

Share your videos with friends, family, and the world

Watch now...

This MSU Codec Comparison is released. It's extremely well done. http://www.compression.ru/video/codec_comparison/codec_comparison_en.html

MSU Video Codecs Comparison (video codec measurement)

MSU Video Codecs Comparison MSU Graphics & Media Lab (Video Group) Project head: Dr. Dmitriy Vatolin Testing, charts, analysis: Sergey Grishin Translating: Daria ...
Read more...

Special Sessions

G.J. Sullivan and Y. Ye, "Recent developments in video coding", Session Chairs, IEEE DCC, Snow Bird, Utah, 4–7 April 2017.

Transcoders

Tr1. E. de la Torre, R.R. Sanchez and J.L. Martinez, "Fast video transcoding from HEVC to VP9", IEEE Trans. on Consumer electronics, vol. 61, pp. 336–343, Aug. 2015. (This has several valuable references on H.264 to HEVC, mpeg-2 to HEVC, and H.264 to SVC transcoders.)

Tr2. E. Peixoto and E. Izquierdo, "A new complexity scalable transcoder from H.264/AVC to the new HEVC codec", IEEE ICIP 2012, session MP.P1.13 (Poster), Orlando, FL, Sept.–Oct. 2012.

Tr3. T. Shanableh, E. Peixoto and E. Izquierdo, "MPEG-2 to HEVC transcoding with content based modeling", IEEE Trans. CSVT, vol. 23, pp. 1191–1196, July 2013.
References [E97] thru [E101] are from VCIP 2012. Pl access http://www.vcip2012.org
There are also some papers on HEVC in the Demo session.

Tr4. E. Peixoto et al., "An H.264/AVC to HEVC video transcoder based on mode mapping", IEEE ICIP 2013, WP.P4.3 (poster), Melbourne, Australia, 15–18 Sept. 2013.

Tr5. E. Peixoto, T. Shanableh and E. Izquierdo, "H.264/AVC to HEVC video transcoder based on dynamic thresholding and content modeling", IEEE Trans. on CSVT, vol. 24, pp. 99–112, Jan. 2014.

Tr6. S. J. Kaware and S. K. Jagtap, "A Survey: Heterogeneous Video Transcoder for H.264/AVC to HEVC", IEEE International Conference on Pervasive Computing (ICPC), pp. 1–3, Pune, India, 8–10 Jan. 2015.

Tr7. E. Peixoto et al., "Fast H.264/AVC to HEVC Transcoding based on Machine Learning", IEEE International Telecommunications Symposium (ITS), pp. 1–4, 2014.

Tr8. Y. Chen et al., "Efficient Software H.264/AVC to HEVC Transcoding on Distributed Multicore Processors," IEEE Trans. on CSVT, vol. 25, pp. 1423–1434, Aug. 2015.

Tr9. A. J. D. – Honrubia et al., "Adaptive Fast Quadtree Level Decision Algorithm for H.264 to HEVC Video Transcoding," IEEE Trans. CSVT, vol. 26, pp. 154–168, Jan. 2016.

Tr10. I. Ahmad, X. Wei, Y. Sun and Y. Zhang, "Video transcoding: an overview of various techniques and research issues," IEEE Trans. Multimedia, vol. 7, pp. 793–804, 2005.

Tr11. A. J. D.-Honrubia et al., "A fast splitting algorithm for an H.264/AVC to HEVC intra video transcoder", IEEE DCC, Snow Bird, Utah, March–April 2016.

Tr12. D. Zhang et al., "Fast Transcoding from H.264 AVC to High Efficiency Video Coding," in IEEE ICME 2012, pp. 651–656, July 2012.

Tr13. D. Guang et al., "Improved prediction estimation based h.264 to HEVC intra transcoding," in Advances in Multimedia Information Processing PCM 2014, ser. Lecture Notes in Computer Science, vol. 8879. Springer International Publishing, pp. 64–73, 2014.

Tr14. D. Hingole, "H.265 (HEVC) bitstream to H.264 (MPEG 4 AVC) bitstream transcoder", M.S. Thesis, EE Dept., University of Texas at Arlington, Arlington, Texas, Dec. 2015.
http://www.uta.edu/faculty/krrao/dip click on courses and then click on EE5359 Scroll down and go to Thesis/Project Title and click on D. Hingole.

Tr15. M. Sugano et al., "An efficient transcoding from MPEG-2 to MPEG-1", IEEE ICIP 2001.

Tr16. N. Feamster and S. Wee, "An MPEG-2 to H.263 transcoder", SPIE conference on multimedia systems and applications II, vol. 3845, pp. 164–175, Sept. 1999.

Tr17. X. lu et al., "Fast mode decision and motion estimation for H.264 with a focus on MPEG-2/H.264 transcoding", IEEE ISCAS, pp. 1246–1249, Kobe, Japan, May 2005.

Tr18. M. Pantoja, H. Kalva and J.B. Lee, "P-frame transcoding in VC-1 to H.264 transcoders", IEEE ICIP, pp. V-297–V-300, 2007. Also several papers on transcoders.

Encryption of HEVC Bit Streams

EH1 B. Boyadjis et al., "Extended selective encryption of H.264/AVC (CABAC) and HEVC encoded video streams", IEEE Trans. CSVT, (early access)

EH-P1 B. Boyadjis et al. [EH1], by incorporating selective encryption of CABAC in regular mode have improved the encryption efficiency over the by-pass mode (see Table VI and Figure 11). In Section V (conclusion and future work), the authors suggest some further studies. Explore and implement future work.

File Format

FF1. M. M. Hannuksela, J. Lainema and V. K. M. Vadakital, "The high efficiency image file format standards," IEEE Signal Processing Magazine, vol. 32, pp. 150–156, July 2015.

FF2. A draft HEIF standard (03/2015) is available at http://mpeg.chiariglione.org/standards/mpeg-h/image-file-format/draft-text-isoiec-fdis-23008-12-carriagestill-image-and-image and the final HEIF standard (ISO/IEC 23008-12) is likely to appear among public ISO standards at http://standards.iso.org/ittf/PubliclyAvailableStandards/.

FF3. The JPEG, Exif, PNG, and GIF specifications are available at http://www.w3.org/Graphics/JPEG/itu-t81.pdf, http://www.cipa.jp/std/documents/e/DC-008-2012_E.pdf, http://www.w3.org/TR/PNG/, and http://www.w3.org/Graphics/GIF/spec-gif89a.txt,respectively.

Online Courses (OLC)

OLC1. Video Lecture on Digital Voice and Picture Communication by Prof. S. Sengupta, Department of Electronics and Electrical Communication Engineering IIT Kharagpur -> https://www.youtube.com/watch?v=Tm4C2ZFd3zE

Open Source Software

OSS1. HEVC open source software (encoder/decoder): https://hevc.hhi.fraunhofer.de/svn/svn_HEVCSoftware/tags/HM-16.9
JCT-VC (2014) Subversion repository for the HEVC test model reference software:
https://hevc.hhi.fraunhofer.de/svn/svn_HEVCSoftware/
http://hevc.kw.bbc.co.uk/svn/jctvc-al24 (mirror site)

OSS2. SHM – 7.0 SHVC reference software:
https://hevc.hhi.fraunhofer.de/svn/svn_SHVCSoftware/tags/SHM-7.0 (See V. Seregin and Y. He, "Common SHM test conditions and software reference configuration", document JCT-VC-P1019, San Jose, CA, Jan. 2014.

OSS3. JSVM 9.19.15
http://ftp3.itu.ch/av-arch/jvt-site/2005_07_Poznan/JVT-P001.doc

OSS4. JCT-3V. (2015, May) MV – and 3D-HEVC reference software, HTM – 14.1 [Online]. Available:
https://hevc.hhi.fraunhofer.de/svn/svn_3DVCsoftware/tags/HTM-14.1/

OSS5. JVT KTA reference software
http://iphome.hhi.de/suehring/tml/download/KTA/

OSS6. JM11KTA2.3
(http://www.h265.net/2009/04/kta-software-jm11kta23.html).

OSS7. JSVM – joint scalable video model reference software for scalable video coding. (online)

http://ip.hhi.de/imagecom_GI/savce/downloads/SVC-Reference-software.htm

OSS8. T. Ritchter: "JPEG XT Reference Codec 1.31 (ISO License)," available online at http://www.jpeg.org/jpegxt/software.html (retrieved July 2015)
JPEG XT: A backwards-compatible extension of JPEG for intermediate and high-dynamic range image compression.
Abstract: JPEG XT is an extension of the well-known JPEG standard that allows lossy or lossless coding of intermediate (8-16 bit integer) or high-dynamic range (floating point) images such that any legacy JPEG decoder is still able to reconstruct an 8-bit/component image from it.
In this talk, the JPEG XT coding technology will be introduced, parts and profiles of the standard will be discussed and a short discussion of the coding performance of the JPEG XT profiles in comparison with other HDR coding technologies will be given.

OSS9. Dirac software: http://sourceforge.net/projects/dirac/

OSS10. AVS-china software download:
ftp://159.226.42.57/public/avs_doc/avs_software

OSS11. HEVC SCC Extension Reference software
https://hevc.hhi.fraunhofer.de/svn/svn_HEVCSoftware/tags/HM-15.0+RExt-8.0+SCM-2.0/

OSS12. Kakadu Software 6.0. [Online].
Available: http://www.kakadusoftware.com/ (JPEG 2000)

OSS13. FFmpeg: Open Source and Cross-Platform Multimedia Library. [Online].
Available: http://www.ffmpeg.org, accessed July 2014. Also https://ffmpeg.org

OSS14. SHVC Reference Software Model (SHM). [Online]. Available: https://hevc.hhi.fraunhofer.de/svn/svn_SHVCSoftware/, accessed July 2014.

OSS15. 3H HEVC Reference Software Model (HTM). [Online]. Available: https://hevc.hhi.fraunhofer.de/svn/svn_3DVCSoftware/, accessed Jul. 2014.

OSS16. Open Source HEVC Decoder (Open HEVC). [Online]. Available: https://github.com/OpenHEVC, accessed July 2014.
OSS16 Kvazaar ultravideo.cs.tut.fi

OSS17. F265 f265.org

OSS18. CES265 ces.itec.kit.edu/ces265.php

OSS19. M. Masera, Adaptive Approximated DCT Architectures for HEVC, 2015 Dec. [Online]. Available: http://personal.det.polito.it/maurizio.masera/material/HEVCSoftware.zip
This software implements the HEVC using these DCT architectures.
OSS20. T.835 : Information technology – JPEG XR image coding system – Reference software," *ITU [Online]*, Available: https://www.itu.int/rec/TREC-T.835-201201-I/.
OSS21. libde open source www.libde265.org
OSS22. Kvazaar HEVC encoder https://github.com/ultravideo/kvzaar open source code supported platforms: x86 and x64 on Windows and Linux
OSS23. cclxv https://bitbucket.org/prunedtree/cclxv OSS HEVC decoder
OSS24. libde265 https://github.com/strukturag/libde265 OSS HEVC decoder

http://x265.ru/en/ – HEVC software
http://www.3gpp.org/DynaReport/26410.htm – HE AAC V2 software
https://ffmpeg.org/download.htmlhttps://ffmpeg.org/download.html - ffmpeg software
http://mkvtoolnix.en.softonic.com/ - mkvmerge software
This MSU Codec Comparison is released. October 2015.
http://www.compression.ru/video/codec_comparison/codec_comparison_en.html
MSU Video Codecs Comparison (video codec measurement)
MSU Video Codecs Comparison MSU Graphics & Media Lab (Video Group) Project head: Dr. Dmitriy Vatolin Testing, charts, analysis: Sergey Grishin Translating: Daria

X265 Source Code

x265 HEVC High Efficiency Video Coding H.265 Encoder. [Online] Available: http://x265.org/, accessed May 6, 2014
http://appleinsider.com/articles/14/09/12/apples-iphone-6-iphone-6-plus-use-h265-codec-for-facetime-over-cellular
Apple's iPhone6, 6plus use H.265 CODEC for encode/decode FaceTime video over cellular. It seems they use A8SoC for this.
Mediatek (MT6595) developed mobile chip H.265 encoder.
VideoLAN x265 HEVC video encoder (online) http://hg.videolan.org/x265

vTune Amplifier by Intel

https://registrationcenter.intel.com/en/forms/?licensetype=2&productid=2489 – This is the direct link for vtune for students to register and download the product for free.

https://registrationcenter.intel.com/en/forms/?licensetype=2&programid=EDUCT&productid=2520 – This is the direct link for vtune for faculty

https://software.intel.com/en-us/qualify-for-free-software – This is the home page link for all Intel software that are free for students, faculty and researchers.

Note: VTune is a bit hidden on this page. Look for Intel Parallel Studio, Cluster Edition. This is a bundle of many tools that includes VTune, compilers, libraries and cluster tools.

General

General1. T. Ebrahimi and M. Kunt, "Visual data compression for multimedia applications," Proc. IEEE, vol. 86, pp. 1109–1125, June 1998.

General2. T. Wiegand and B. Girod, "Lagrange multiplier selection in hybrid video coder control," in Proc. IEEE ICIP, vol. 3, pp. 542–545, Oct. 2001.

General3. A.M. Patino et al., "2D-DCT on FPGA by polynomial transformation in two-dimensions," Proceedings of the IEEE Intnl. Symp. on Circuits and Systems, ISCAS '04, vol. 3, pp. 365–368, May 2004.

General4. G.A. Davidson et al., "ATSC video and audio coding," *Proc. of the IEEE*, vol. 94, pp. 60–76, Jan. 2006.

General5. J. Golston and A. Rao, "Video compression: System trade-offs with H.264, VC-1 and other advanced CODECs," Texas Instruments, White Paper, Nov. 2006. This paper was written for and presented at the Embedded Systems Conference – Silicon Valley, San Jose, CA, April 2006.

General6. C.-H. Yu and S.-Y. Chen, "Universal colour quantisation for different colour spaces," IEE Proc. Vision, Image and Signal Processing, vol. 153, pp. 445–455, Aug. 2006.

General7. R.A. Burger et al., "A survey of digital TV standards China," IEEE Second Int'l. Conf. Commun. and Networking in China, pp. 687–696, Aug. 2007.

General8. T. Wiegand and G.J. Sullivan, "The picture phone is here. Really," IEEE Spectrum, vol. 48, pp. 50–54, Sept. 2011.

General9. N. Jayant, "Frontiers of audiovisual communications: New convergences of broadband communications, computing, and rich media," Proc. IEEE, Special Issue, Scanning the Issue, vol. 100, pp. 821–823, April 2012. Several papers on audio visual communications.

General10. H.R. Wu et al., "Perceptual visual signal compression and transmission", Proc. IEEE, vol. 101, pp. 2025–2043, Sept. 2013.

General11. M.A. Isnardi, "Historical Overview of video compression in consumer electronic devices", IEEE ICCE, pp. 1–2, Las Vegas, NV, Jan. 2007.

General12. G. J. Sullivan and T. Wiegand, "Video compression – from concepts to the H.264/AVC Standard," Proc. IEEE, vol. 93, pp. 18–31, Jan. 2005.

General13. G. J. Sullivan et al., "Future of Video Coding and Transmission," IEEE Signal Processing Magazine, Vol. 23, pp. 76–82, Nov. 2006.

General14. T. Ebrahimi and M. Kunt, "Visual Data Compression for Multimedia Applications," Proc. IEEE, vol. 86, pp. 1109–1125, June 1998.

General15. Proc. IEEE, Special Issue on Global Digital Television: Technology and Emerging Services, vol. 94, Jan. 2006.

General16. N. Jayant, J. Johnston, and R. Safranek, "Signal compression based on models of human perception," Proc. IEEE, vol. 81, pp. 1385–1422, Oct. 1993.

General17 R. Schafer and T. Sikora, "Digital video coding standards and their role in video communications", Proc. IEEE, vol. 83, pp. 907–924, June 1995.

General18 A.K. Jain, "Image data compression: A review", Proc. IEEE, vol. 69, pp. 349–384, March 1981.

JVT REFLECTOR Queries/questions/clarifications etc. regarding H.264/H.265

jvt-experts-bounces@lists.rwth-aachen.de; on behalf of; Karsten Suehring [karsten.suehring@hhi.fraunhofer.de]

The JCT experts reflector is used for discussions related to H.264/AVC and HEVC. You can subscribe yourself here: http://mailman.rwth-aachen.de/mailman/listinfo/jvt-experts

For discussions related to HEVC you should use the JCT-VC email reflector that can be found here:
http://mailman.rwth-aachen.de/mailman/listinfo/jct-vc
The draft standard: High Efficiency Video Coding Draft 10, Document JCTVC-L1003, available at:
http://phenix.it-sudparis.eu/jct/doc_end_user/current_document.php?id=7243
JVT KTA reference software:
http://iphome.hhi.de/suehring/tml/download/KTA/

HEVC Quality Evaluation

http://www.slideshare.net/touradj_ebrahimi/subjective-quality-evaluation-of-the-upcoming-hevcvideo-compression-standard
Fraunhofer HHI Institute, Germany: Image processing research group
http://www.hhi.fraunhofer.de/en/fields-of-competence/image-processing/research-groups/image-video-coding/team.html

Video Test Sequences

Repository for freely-redistributable test sequences. [Online] Available: www.media.xiph.org
"Test video sequences." [Online]. Available:
ftp://hvc:US88Hula@ftp.tnt.uni-hannover.de/testsequences

References on SSIM

SSIM1. Z. Wang, et al., "Image quality assessment: From error visibility to structural similarity," IEEE Trans. Image Processing, vol. 13, pp. 600–612, Apr. 2004.
SSIM2. Z. Wang, L. Lu and A. C. Bovik, "Video quality assessment using structural distortion measurement", IEEE International Conference on Image Processing, Vol. 3, pp. 65–68, Rochester, NY, Sep. 22–25, 2002.
SSIM3. The SSIM Index for Image Quality Assessment
http://www.ece.uwaterloo.ca/~z70wang/research/ssim/
SSIM4. What is 3-SSIM? 2/8/11 Alexander Parshin [videocodec-testing@graphics.cs.msu.ru]
It is our implementation of
http://live.ece.utexas.edu/publications/2010/li_jei_jan10.pdf
SSIM5. W. Malpica and A. Bovik, "Range image quality assessment by structural similarity", IEEE ICASSP 2009, pp. 1149–1152, 19–24 April 2009.

SSIM6. J. Zujovic, T.N. Pappas and D.L. Neuhoff, "Structural similarity metrics for texture analysis and retrieval", IEEE ICIP 2009, Cairo, Egypt, Nov. 2009 (paper MP.L1.5)

SSIM7. X. Shang, "Structural similarity based image quality assessment: pooling strategies and applications to image compression and digit recognition," M.S. Thesis, EE Department, The University of Texas at Arlington, Aug. 2006.

SSIM8. Z. Wang and Q. Li, "Video quality assessment using a statistical model of human visual speed perception", *Journal of the Optical Society of America A,* vol. 24, no. 12, pp. B61–B69, Dec. 2007.

SSIM9. Z. Wang and X. Shang, "Spatial pooling strategies for perceptual image quality assessment," *IEEE International Conference on Image Processing*, pp. 2945–2948, Atlanta, GA, Oct. 8–11, 2006.

SSIM10. Z. Wang and E. P. Simoncelli, "Translation insensitive image similarity in complex wavelet domain," *IEEE International Conference on Acoustics, Speech and Signal Processing*, vol. II, pp. 573–576, Philadelphia, PA, Mar. 2005.

SSIM11. Z. Wang, L. Lu, and A. C. Bovik, "Video quality assessment based on structural distortion measurement," *Signal Processing: Image Communication*, special issue on "Objective video quality metrics", vol. 19, no. 2, pp. 121–132, Feb. 2004.

SSIM12. Z. Wang, E. P. Simoncelli and A. C. Bovik, "Multi-scale structural similarity for image quality assessment," Invited Paper, *IEEE 37th Asilomar Conference on Signals, Systems and Computers*, vol. 2, pp. 1398–1402, Nov. 2003.
Dr. Zhou Wang's web site
http://www.ece.uwaterloo.ca/~z70wang/temp/DIP.zip
Anush K. Moorthy and Alan C. Bovik
Efficient motion weighted spatio-temporal video SSIM index
Proc. SPIE, Vol. 7527, 75271I (2010); doi:10.1117/12.844198
Conference Date: Monday 18 January 2010 Conference Location: San Jose, California, USA
Conference Title: Human Vision and Electronic Imaging XV

SSIM13. S. Wang, S. Ma and W. Gao, "SSIM based perceptual distortion rate optimization coding", SPIE, VCIP, vol. 7744–91, Huangshan, China, July 2010.
Is & T Electronic Imaging, Science and Technology, SPIE, Image quality and system performance, vol. 7867, San Francisco, CA, Jan. 2011. (plenary speech by A. C. Bovik).

SSIM14. A. Bhat, I. Richardson and S. Kanangara, "A new perceptual quality metric for compressed video based on mean squared error", SP:IC, pp. 1–4, 30 July, 2010.

SSIM15. Y.-T. Chen et al., "Evaluation of video quality by CWSSIM method", SPIE, Mathematics of Data/Image Coding, Compression, and Encryption with Applications XII, vol. 7799, pp. 7799T – 1–7, Aug. 2010.

SSIM16. J. Wang et al., "Fractal image coding using SSIM", IEEE ICIP 2011.

SSIM17. A. Rehman and Z. Wang, "SSIM based non-local means image denoising", IEEE ICIP 2011.

SSIM18. W. Lin and C.-C. J. Kuo, "Perceptual visual quality metrics: A survey", J. VCIR, vol. 22, pp. 297–312, May 2011.

SSIM19. C. Vu and S. Deshpande, "ViMSSIM: from image to video quality assessment", ACMMoVid 12, Proc. 4th workshop on mobile video, Chapel Hill, N.C., Feb. 2012.

SSIM20. A. Horne and D. Ziou, "Image quality metrics: PSNR vs. SSIM", IEEE ICPR, pp. 2366–2369, 2010.

SSIM21. C. Yeo, H.L. Tan and Y.H. Tan, "SSIM-based adaptive quantization in HEVC", IEEE ICASSP, pp. 1690–1694, Vancouver, Canada, 2013. (IVMSP-P3.2: Video coding II Poster).
"SSIM-based Error-resilient Rate Distortion Optimized H.264/ AVC Video Coding for Wireless Streaming" Signal Processing: Image Communication, pp. 2332–2335, June 2013.

SSIM22. T.-S. Ou and Y.-H. Huang and H.H. Chen, "SSIM-based perceptual rate control for video coding", IEEE Trans. CSVT, vol. 21, pp. 682–691, May 2011.

SSIM23. A. Rehman and Z. Wang, "SSIM-inspired perceptual video coding for HEVC", IEEE ICME 2012, pp. 497–502, July 2012. **(Same as SSIM29)**.

SSIM23. S. Wang et al., "SSIM-motivated rate distortion optimization for video coding", IEEE Trans. CSVT, vol. 22, pp. 516–529, April 2012.

SSIM24. C. Yeo, H.L. Tan and Y.H. Tan, "On rate distortion optimization using SSIM", IEEE Trans. on CSVT, vol. 23, pp. 1170–1181, July 2013.

SSIM25. T. Zhou and Z. Wang, "On the use of SSIM in HEVC", IEEE Asilomar conf. on circuits, systems and computers, pp. 1107–1111, Nov. 2013.

SSIM26. M. Hassan and C. Bhagvati, "Structural similarity measure for color images," Int J Comput Appl, vol. 43, pp. 7–12, 2012.

SSIM27. K. Naser, V. Ricordel and P. L. Callet, "Experimenting texture similarity metric STSIM for intra prediction mode selection and block partitioning in HEVC", DSP, pp. 882–887, 2014.

SSIM28. S. Wang et al., "Perceptual Video Coding Based on SSIM – Inspired Divisive Normalization," IEEE Trans. on Image Processing, vol. 22, no. 4, pp. 1418–1429, Apr. 2013.

SSIM29. A. Rehman and Z. Wang, "SSIM – Inspired Perceptual Video Coding for HEVC," Proc. Int. Conf. Multimedia and Expo, pp. 497–502, Melbourne, Australia, July. 2012.

SSIM30. C.L. Yang, "An SSIM-Optimal H.264/AVC Inter Frame Encoder", Proceedings of 8th IEEE/ACIS International Conference on Computer and Information Science (ICIS 2009), pp. 291–295, Shanghai, China, June 2009.

SSIM31. G.H. Chen, "Edge-based Structural Similarity for Image Quality Assessment", IEEE ICASSP, vol. 2, pp. 933–936, Toulouse, France, May 2006.

SSIM32. M.-J. Chen and A. C. Bovik, "Fast structural similarity index algorithm", Journal of Real-Time Image Processing, vol. 6, no. 4, pp. 281–287, Dec. 2011.

SSIM33. W.-T. Loh and D.B.L. Bong, "Temporal video quality assessment method involving structural similarity", IEEE-ICCE Taiwan, May 2016.

SSIM34. S. Wang et al., SSIM-motivated two-pass VBR video coding for HEVC", IEEE Trans. CSVT (early access). This has several references related to SSIM.
Emmy award for SSIM (Oct. 2015)
http://www.emmys.com/news/awards-news/sthash.c1DcwJf7.dpuf
Zhou Wang, Alan Bovik, Hamid Shiekh and Eero Simoncelli for Structural Similarity (SSIM) Video Quality Measurement Model. Structural Similarity (SSIM) is an algorithm for estimating the perceived quality of an image or video. Its computational simplicity and ability to accurately predict human assessment of visual quality has made it a standard tool in broadcast and post-production houses throughout the television industry.

SSIM uses powerful neuroscience-based models of the human visual system to achieve breakthrough quality prediction performance. Unlike previous complex error models that required special hardware, it can be easily applied in real time on common processor software. SSIM is now the most widely-used perceptual video quality measure, used to test and refine video quality throughout the global cable and satellite TV industry. It directly affects the viewing experiences of tens of millions of viewers daily.
SSIM by Zhou wang on You tube
Objective image quality assessment, what's beyond_
https://www.youtube.com/watch?v=ibiCs_NJgCQ

Bjontegaard Metric

BD1 G. Bjontegaard, "Calculation of average PSNR differences between RD-Curves", ITU-T SG16, Doc. VCEG-M33, 13th VCEG meeting, Austin, TX, April 2001. http://wfpt3.itu.int/av-arch/video-site/0104_Aus/VCEG-M33.doc
A graphical explanation of BD-PSNR and BS-bitrate is shown in Tabatabai et al. [E202].
BD2 G. Bjontegaard, "Improvements of the BD-PSNR model", ITU-T SG16 Q.6, Doc. VCEG-AI11, Berlin, Germany, July 2009.
BD3 K. Anderson, R. Sjobetg and A. Norkin, "BD measurements based on MOS", (online), ITU-T Q6/SG16, document VCEG-AL23, Geneva, Switzerland, July 2009. Available: http://wfpt3.itu.int/av-arch/video-site/0906_LG/VCEG-AL23.zip
BD4 S. Pateux and J. Jung, "An Excel add-in for computing Bjontegaard metric and its evolution," VCEG Meeting, Document VCEG – AE07, Marrakech, Morocco, Jan. 2007.
BD5 F. Bossen, "Excel template for BD-rate calculation based on piece-wise cubic Interpolation", JCT-VC Reflector, 2011.

BD Metrics computation using MATLAB source code [Online] Available:
http://www.mathworks.com/matlabcentral/fileexchange/41749-bjontegaard-metric-calculation–bd-psnr-/content/bjontegaard2.m BD Metrics: VCEG-M34. [Online] Available: http://wftp3.itu.int/av-arch/video-site/0104_Aus/VCEG-M34.xls

See also
M. Wien, "High Efficiency Video Coding: Coding Tools and Specification", Springer, 2014. Section 3.6.1.3 pages 96–97.
Figure below explains clearly the BD-bit rate and BD-PSNR [E202].

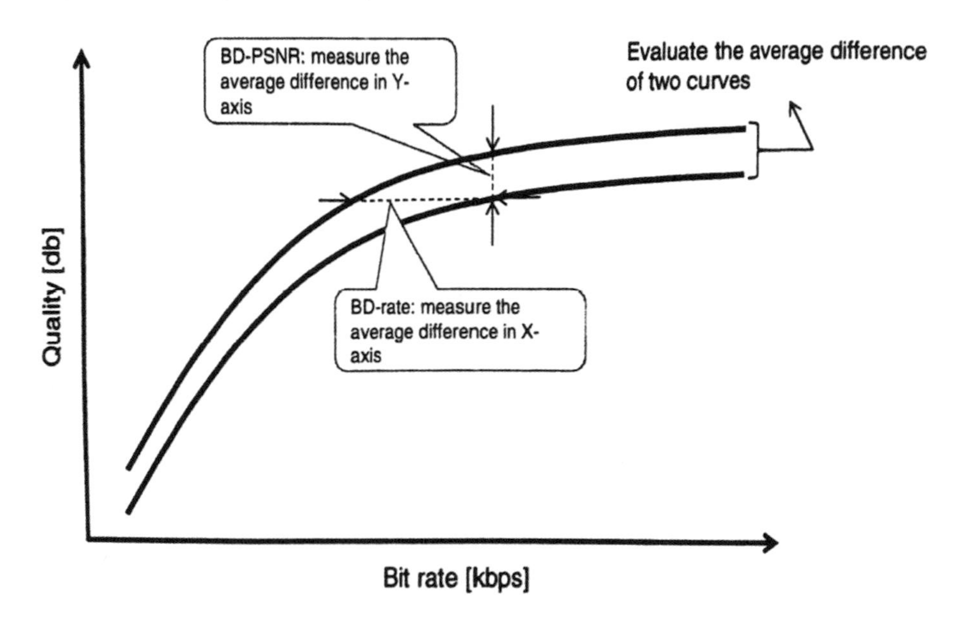

VP8, VP9, VP10

VP.1. D. Grois, et al., "Performance comparison of H.265/MPEG-HEVC, VP9, and H.264/MPEGAVC encoders," in Proc. 30th IEEE Picture Coding Symposium, pp. 394–397, Dec. 2013.

VP.2. A. Grange. *Overview of VP-Next*. [Online]. Available: http://www.ietf.org/proceedings/85/slides/slides-85-videocodec-4.pdf, accessed Jan. 30, 2013.

VP.3. J. Bankoski, P. Wilkins, and Y. Xu, "Technical overview of VP8, an open source video codec for the web," in *Proc. IEEE Int. Conf. Multimedia* and *Expo*, pp. 1–6, Jul. 2011.

VP.4. M. Rerabek and T. Ebrahimi, "Comparison of compression efficiency between HEVC/H.265 and VP9 based on subjective assessments", SPIE Optical Engineering + Applications, vol. 9217, San Diego, CA, Aug. 2014.

VP.5. D. Mukherjee et al., "An overview of new video coding tools under consideration for VP10: the successor to VP9," [9599 – 50], SPIE. Optics + photonics, Applications of digital image processing, XXXVIII, San Diego, California, USA, 9–13, Aug. 2015. Debargha Mukherjee email: debargha@google.com
Please see the last slide below:
Conclusions

- Modest progress towards a next generation codec
- A long way to go still
- Need a few big ideas
- Welcome to join the VP10 effort!
- All development in the open.
- Join the discussion with fellow developers
- The mailing list and group information can be found at http://www.webmproject.org/about/discuss/

VP.6. D. Mukherjee et al., "The latest open – source video codec VP9 – an overview and preliminary results," Proc. IEEE picture coding Symposium, pp. 390–393, San Jose, CA, Dec. 2013.

VP.7. D. Mukherjee et al., "A Technical overview of VP9 – the latest open – source video codec," SMPTE Motion Imaging Journal, Jan/Feb 2015.

VP.8. E. Ohwovoriole, and Y. Andreopoulos, "Rate-Distortion performance of contemporary video codecs: comparison of Google/WebM VP8, AVC/H.264 and HEVC TMuC", Proc. London Communications Symposium (LCS), pp. 1–4, Sept. 2010.

VP.9. J. Bankoski, P. Wilkins and X. Yaowu, "Technical overview of VP8, an open source video codec for the web", IEEE International Conference on Multimedia and Expo (ICME), pp. 1–6, 11–15 July 2011.

VP.10. Chromium® open-source browser project, VP9 source code, Online: http://git.chromium.org/gitweb/?p=webm/libvpx.git;a=tree;f=vp9; hb=aaf61dfbcab414bfacc3171 501be17d191ff8506

VP.11. J. Bankoski, et al., "Towards a next generation open-source video codec," Proc. SPIE 8666, Visual Information Processing and Communication IV, pp. 1–13, Feb. 21, 2013.

VP.12. E. de la Torre, R.R. Sanchez and J.L. Martinez, "Fast video transcoding from HEVC to VP9", IEEE Trans. on Consumer electronics, vol. 61, pp. 336–343, Aug. 2015. (This has several valuable references on H.264 to HEVC, MPEG-2 to HEVC, and H.264 to SVC transcoders.) also "HEVC to VP9 transcoder", IEEE VCIP 2015.

VP.13 See J. Padia, "Complexity reduction for VP6 to H.264 transcoder using motion vector reuse", M.S. Thesis, EE Dept., University of Texas at Arlington, Arlington, Texas, May 2010. http://www.uta.edu/faculty/krrao/dip click on courses and then click on EE5359 Scroll down and go to Thesis/Project Title and click on Jay Padia. Develop similar technique for VP9 to HEVC transcoder.

VP.14 "VP9 video codec", https://www.webmproject.org/vp9

VP.15 G. Paim et al., "An efficient sub-sample interpolator hardware for VP9–10 standards", IEEE ICIP, Phoenix, Arizona, Sept. 2016.

VP.16 D. Mukherjee, et al., "A technical overview of VP9 – the latest open-source video codec", *SMPTE*, vol. 2013, no. 10, pp. 1–17, Annual technical conference and exhibition, 22–24 Oct. 2013.

VP-P1 By cleverly using the information from the decoding process (HEVC) to accelerate the encoding process (VP9), the authors [VP12] have achieved significant reduction in encoding complexity (VP9) with negligible loss in PSNR. However BD bit rate, BD PSNR and SSIM metrics have not been used. Include these metrics in evaluating the transcoder.

VP-P2 See VP5. VP10 being developed as next generation codec by Google and successor to VP9 is an open source royalty free codec. VP9 is currently being served extensively by You Tube resulting in billions of views daily. This paper describes several tools that are proposed beyond VP9 with the objective of improving the compression ratio (reduced bit rate) while preserving the same visual quality as VP9. (This is similar to development of H.265/HEVC in comparison with H.264/AVC). VP10 project is in still early stages and Google welcomes proposals in its development from researchers and developers. The tools (listed below) being evaluated in [VP.5] are subject to change including the exploration of new tools. Implement each of these tools separately and evaluate their performance compared with baseline codec VP9. Use the three test sets as described under Section 3 coding results. Please confirm the improvements caused by these tools over VP9 as described in Tables 1 through 5. Implementation complexity caused by these tools needs also to be considered as another performance metric. Evaluate this in comparison with VP9. Explore/innovate additional tools that can further improve the coding efficiency of VP10. The authors state that VP10 development is an open – source project, and they invite the rest of the video coding community to join the effort to create tomorrow's royalty free codec.

- New coding tools, new ways of combining existing/new methods
- Several experiments and investigative threads underway
- High bit-depth internal
- Prediction tools
- Transform coding tools
- Screen content coding
- Miscellaneous

VP-P3 See VP.4 and VP.8 Performance comparison of HEVC, VP9 and H.264/MPEG 4 AVC in terms of subjective evaluations showing actual differences between encoding algorithms in terms of perceived quality is implemented. The authors indicate a dominance of HEVC based encoding algorithms compared to both VP9 and H.264/MPEG 4 AVC over wide range of bit rates. Extend this comparison in terms of PSNR, BD-bit rates, BD-PSNR, SSIM and implementation complexity. Consider various test sequences at different spatial and temporal resolutions.

VP-P4 Performance comparison of SDCT with DCT and DDCT is investigated in [E392]. In VP10 INTDCT is used. Replace INTDCT with INTSDCT and implement the performance comparison based on various test sequences including 4K and 8K and different block sizes. See the Tables and Figures in [E392].

Integrating inside VP10 might require a significant amount of work, as the transform has to be inserted in the rate-distortion optimization loop, and auxiliary information may have to be signaled. What can be easily done is to take the VP10 integer transform and rotate the transform using the technique developed in [E392] thus obtaining a rotated transform.

VP-P5 See [E393]. Apply the FCDR to VP10 codec both as a post processing operation and as an in loop (embedded) operation. Implement the projects for VP10 similar to those described in P.5.268 and P.5.269. The authors state that the merging of plenoptic imaging and correlation quantum imaging has thus the potential to open a totally new line of research. Explore this.

JPEG 2000

J2K1. D. S. Taubman and M. W. Marcellin, "JPEG 2000: Image compression fundamentals, standards and practice," Springer, 2002.

J2K2. M. Rabbani, Review of the book: (D. S. Taubman and M. W. Marcellin, "JPEG 2000: Image compression fundamentals, standards and practice," Springer, 2002), *J. Electron. Imaging,* vol. 11, no. 2, p. 286, Apr. 2002.

J2K3. Kakadu Software 6.0. [Online]. Available: http://www.kakadusoft ware.com/

J2K4. H. Oh, A. Bilgin and M. W. Marcellin, "Visually lossless encoding for JPEG 2000," IEEE Trans. on Image Processing, vol. 22, pp. 189–201, Jan. 2013.

J2K5. C. Christopoulus, A. Skodras and T. Ebrahimi, "The JPEG 2000 still image coding system: An overview," IEEE Trans. on Consumer Electronics, vol. 46, pp. 1103–1127, Nov. 2000.

J2K6. J. Hunter and M. Wylie, "JPEG 2000 Image Compression: A real time processing challenge," Advanced Imaging, vol. 18, pp. 14–17, April 2003.

J2K7. D. Marpe, V. George and T. Wiegand, "Performance comparison of intra – only H.264/AVC HP and JPEG 2000 for a set of monochrome ISO/IEC test images," JVT – M014, pp. 18–22, Oct. 2004.

J2K8. D. Marpe et al., "Performance evaluation of motion JPEG 2000 in comparison with H.264/operated in intra – coding mode, "Proc. SPIE, vol. 5266, pp. 129–137, Feb 2004.

J2K9. JPEG2000 latest reference software (Jasper Version 1.900.0) Website: http://www.ece.uvic.ca/~frodo/jasper/

J2K10. M.D. Adams, "JasPer software reference manual (Version 1.900.0)," ISO/IEC JTC 1/SC 29/WG 1 N 2415, Dec. 2007.

J2K11. M.D. Adams and F. Kossentini, "Jasper: A software-based JPEG-2000 codec implementation," in *Proc. of IEEE Int. Conf. Image Processing*, vol. 2, pp. 53–56, Vancouver, BC, Canada, Oct. 2000.

J2K12. C. Christopoulos, J. Askelof and M. Larsson, "Efficient methods for encoding regions of interest in the upcoming JPEG2000 still image coding standard", IEEE Signal Processing Letters, vol. 7, pp. 247–249, Sept. 2000.

J2K13. F. Dufaux, "JPEG 2000 – Extensions", ITU-T VICA Workshop ITU Headquarters – Geneva, 22–23 July, 2005. (PPT slides)

J2K14. T. Fukuhara et al., "Motion-JPEG2000 standardization and target market, IEEE ICIP, vol. 2, pp. 57–60, 2000.

J2K15. M. Rabbani and R. Joshi, "An overview of the JPEG 2000 still image compression standard," Signal Processing: Image Communication, vol. 17, pp. 3–48, Jan. 2002.

J2K16. J. Hunter and M. Wylie, "JPEG2000 Image Compression: A real time processing challenge," Advanced Imaging, vol. 18, pp. 14–17, and 43, April 2003.

J2K17. The JPEG 2000 Suite" (Peter Schelkens, editor; Athanassios Skodras and Touradj Ebrahimi, co-editors)—an addition to the IS&T/Wiley Series, 2010.

J2K18. P. Topiwala, "Comparative Study of JPEG2000 and H.264/AVC FRExt I– Frame Coding on High-Definition Video Sequences" SPIE, vol. 5909-31, Sept. 2005.

J2K19. T.D. Tran, L. Liu and P. Topiwala, "Performance Comparison of leading image codecs: H.264/AVC Intra, JPEG2000, and Microsoft HD Photo", http://www.uta.edu/faculty/krrao/dip/Courses/EE5351/AIC7.pdf

J2K20. D.T. Lee, "JPEG2000: Retrospective and new developments", Proc. IEEE, vol. 93, pp. 32–41, Jan. 2005.

J2K21. Special issue on JPEG 2000, Signal Processing and Image Communication, vol. 17, Jan. 2002.

J2K22 C. Li, C. Deng and B. Zhao, "Textural and gradient feature extraction from JPEG2000 code stream for airfield detection", IEEE DCC, Snow Bird, Utah, March–April 2016.

J2K23 D. Barina, O. Klima and P. Zemcik, "Single-loop software architecture for JPEG2000", IEEE DCC, Snow Bird, Utah, March-April 2016.

J2K24 T. Richter and S. Simon, "Towards high-speed, low-complexity image coding: variants and modification of JPEG 2000," in Proc. SPIE, vol. 8499, 2012.

J2K25 F. De Simone et al., "A comparative study of JPEG 2000, AVC/H.264, and HD photo," in SPIE Optics and Photonics, Applications of Digital Image Processing, vol. 6696, 2007.

J2K26 F. Liu et al., "Visibility thresholds in reversible JPEG2000 compression", IEEE DCC, April 2017.

Digital Cinema

The Digital Cinema (DC) Ad Hoc Group within the JPEG Committee has been successful in seeing their work adopted by the industry. The Digital Cinema Initiatives (www.dcimovies.com) organization has adopted JPEG 2000 for the distribution of digital movies to theatres. The successful roll-out of this solution continues unabated with over 5000 theatres supporting digital cinema, including 1000 stereoscopic theatres. The first live-action stereoscopic feature encoded with JPEG 2000, Beowulf, is being released on November 16, 2007. The Digital Cinema Ad Hoc Group has initiated work on studio broadcast applications. In addition, the group has begun work on the archival of motion pictures and related contents.

JPEG 2000 Related Work

At the Kobe meeting, the JPEG Committee confirmed the endorsement of Intellectual Resource Initiative (IRI) as JTC 1 Registration Authority (RA) for use in JPSEC. IRI is a Non Profit Organization, based in Japan, established to create policy proposals on the importance of intellectual information in society.

JPEG 2000 Part 9 known as JPIP, allows powerful and efficient network access to JPEG 2000 images and their metadata in a way that exploits the best features of the JPEG 2000 standard. Interoperability testing among several JPIP implementations continued. Participation from additional organizations is solicited, and testing will continue over the Internet between meetings. For more information please contact jpip@jpeg.org.

JPEG 2000 Part 10 Ad Hoc Group has been working on the extension of JPEG 2000 to three-dimensional images such as Computer Tomography (CT) scans and scientific simulations. JP3D is currently in balloting phase for International Standard (IS) status. Due to an increased interest for compression technologies for floating-point data, the JPEG Committee has issued a call for information on applications and compression technology for floating-point data. Responses will be reviewed at the 44th WG1 San Francisco Meeting, March 31–April 4, 2008.

JPEG 2000 Part 11 Wireless, also known as JPWL, has become an International Standard (ISO/IEC 15444-11). JPWL has standardized tools and methods to achieve the efficient transmission of JPEG 2000 imagery over an error-prone wireless system.

JPEG 2000 Part 13 standardizes an entry level JPEG 2000 encoder with widespread applications, intended to be implemented on a license and royalty fee free basis. It is published as ISO/IEC 15444-13.

JPSearch

ISO/IEC 24800, Still Image Search, known as JPSearch, is a project that aims to develop a standard framework for searching large collections of images. This project is divided in five parts.

- Part 1 – Framework and System Components, is a Technical Report that introduces the JPSearch architecture and outlines the organization of the JPSearch specifications.
- Part 2 – Schema and Ontology Registration and Identification, standardizes a format for the import, export and exchange of ontology.
- Part 3 – JPSearch Query Format, which is developed jointly with MPEG, allows for the expression of search criteria, the aggregation of return results and the management of query process.
- Part 4 – Metadata Embedded in Image Data (JPEG and JPEG 2000) file format, standardizes image data exchange format with associated metadata.
- Part 5 – Data Interchange Format between Image Repositories, standardizes a format for the exchange of image collections and respective metadata between JPSearch compliant repositories.

At the meeting, requirements for each part have been reviewed and updated, and Working Drafts for Part 2 and Part 4 and Committee Draft of Part 3 have been produced.

The new JasPer (JPEG-2000) release is available from the JasPer Project Home Page (i.e., http://www.ece.uvic.ca/∼mdadams/jasper)

JasPer 1.900.21 is available from the JasPer home page: (Nov. 2016) http://www.ece.uvic.ca/∼mdadams/jasper

(i.e., http://www.jpeg.org/software) JasPer is the software in C for JPEG2000. Also http://jj2000.epfl.com

http://www.kakadusoftware.com (kakadu software by David Taubman)

JPEG-2000 JasPer software – http://www.ece.uvic.ca/∼mdadams/jasper

JPEG-2000 tutorial – http://www.ece.uvic.ca/∼mdadams

Dear JasPer Users, 12/30/2016

I am pleased to announce the release of version 2.0.10 of the JasPer software. This new release can be obtained from the“Downloads” section of the JasPer web site:

http://www.ece.uvic.ca/~frodo/jasper/#download
If you encounter any problems with the software, please report them to the issue-tracker for JasPer on GitHub:
https://github.com/mdadams/jasper/issues
—
Michael Adams, Associate Professor, Ph.D., P.Eng., IEEE Senior Member
Dept. of Electrical and Computer Engineering, University of Victoria,
PO Box 1700 STN CSC, Victoria, BC, V8W 2Y2, CANADA
E-mail: mdadams@ece.uvic.ca, Web: http://www.ece.uvic.ca/~mdadams

Sweden Records Missile Tests Using JPEG2000 Technology April 2009

The Swedish Armed Forces deploys a sophisticated, medium-range ground-to-air defense missile system, called RBS 23 BAMSE, manufactured by Swedish company Saab Bofors Dynamics (SBD).

To facilitate the test system, SBD utilizes RGB Spectrum's advanced DGy ultra high resolution recording systems. The DGy™ recorders were selected based upon their superior ability to record and reproduce the most intricate details using advanced JPEG2000 compression. Previously, SBD used scan converters and conventional VCRs which rendered "soft" imagery with difficult to discern text and graphics. The DGy recorders preserve the system's complex symbologies and provide sharp graphics and alpha-numerics for after-action-review.

RGB Spectrum Real Time Newsletter 6/25/09

Advanced JPEG2000 Recording Deployed For Helicopter Mission Simulators

Rockwell Collins, the prime contractor for the U.K. Army's Aviation Command and Tactics Trainer (ACTT), was contracted to execute the sixth phase upgrade of this simulator. The upgrade was designed to provide advanced command and tactical training capability for the Army's helicopter pilots for operation in high threat environments.

The upgrade required a system to record the intricate symbology used in the simulated Integrated Helmet and Display Sight System (IHADSS) unit. Rockwell Collins selected RGB Spectrum's DGy™ digital recording system to meet this demanding requirement. The DGy system was chosen for its superior ability to reproduce intricate detail, achieved through advanced wavelet-based JPEG2000 near lossless compression.

Project

J2K-P1. See references [36] through [39] in [LL1]. These references relate to parallel processing methods and advanced graphic processing units for significant speed up in execution time of JPEG 2000. Review these references and evaluate the encoding and decoding speeds of JPEG 2000. Consider the encoder/decoder complexity reduction using various test images.

J2K-P2. See J2K24. By replacing the EBCOT coding with a simple Huffman-run length scheme, the JPEG 2000 implementation complexity is significantly reduced. Implement JPEG 2000 Huffman coding and evaluate how the simplified VLC reduces the encoder/decoder complexity.

JPEG XR (Extended Range)

JPEG XR

JXR1. F. Dufaux, G. J. Sullivan, and T. Ebrahimi, "The JPEG XR image coding standard [Standards in Nutshell]," IEEE Signal Process. Magazine, vol. 26, no. 6, pp. 195–199 and 204, Nov. 2009.

JXR2. Kodak Lossless True Color Image Suite. [Online]. Available: http://r0k.us/graphics/kodak/

JXR3. JPEG Core Experiment for the Evaluation of JPEG XR Image Coding. [Online]. Available: http://jahia-prod.epfl.ch/site/mmspl/op/edit/page-58334.html

JXR4. Microsoft HD photo specification: http://www.microsoft.com/whdc/xps/wmphotoeula.mspx

JXR5. ITU. (2012) JPEG XR image coding system reference software. [online]. Available: http://www.itu.int/rec/T-REC-T.835-201201-I/en

JXR6 C. Tu et al., "Low-complexity hierarchical lapped transform for lossy-to-lossless image coding in JPEG XR/HD Photo", Proc. SPIE, Vol. 7073, pp. 70730C-1–70730C-12, San Diego, CA, Aug. 2008.

JXR7 T. Suziki and T. Yoshida, " Lower Complexity Lifting Structures for Hierarchical Lapped Transforms Highly Compatible with JPEG XR Standard", IEEE Trans. CSVT (early access). The authors state, "JPEG XR (eXtended Range) is a newer lossy – to – lossless image coding standard. JPEG XR has half the complexity of JPEG 2000 while preserving image quality".

JXR8 T.835 : Information technology – JPEG XR image coding system – Reference software," *ITU [Online]*, Available: https://www.itu.int/rec/TREC-T.835-201201-I/.

JXR9 "JPEG core experiment for the evaluation of JPEG XR image coding," *EPFL, Multimedia Signal Processing Group [Online]*, Available: http://mmspg.epfl.ch/iqa.

JXR10 "The New Test Images – Image Compression Benchmark," *Rawzor-Lossless Compression Software for Camera RAW Images [Online]*, Available: http://imagecompression.info/test images/.

JPEG XR is a proposed new part (Part 2) of the recently established new work item in JPEG, known as JPEG Digital Imaging System Integration (ISO/IEC 29199 – JPEG DI). JPEG DI aims to provide harmonization and integration between a wide range of existing and new image coding schemes, in order to enable the design and delivery of the widest range of imaging applications, across many platforms and technologies. JPEG DI aims to leverage the rich array of tools developed in and around JPEG and JPEG 2000 to support new image compression methods such as JPEG XR. JPEG XR is designed explicitly for the next generation of digital cameras, based extensively on the technology introduced by Microsoft in its Windows Media Format proposals, at present known as HD Photo. At the Kobe meeting, The JPEG XR specification (Working Draft) was reviewed and will be balloted for promotion to Committee Draft (CD) status before the 44th WG1 San Francisco meeting, March 31 to April 4, 2008. In addition to JPEG XR itself, the creation of two other new parts of the standard on compliance testing and reference software was approved at the meeting. Now adopted as a standard (Nov. 2009.)

AIC Advanced Image Coding (latest activity in still frame image coding) please access the web site below.

http://www.bilsen.com/index.htm?http://www.bilsen.com/aic/

Pl access the web site below about ELYSIUM activity in JPEG/JPEG 2000/JPEG XR and AIC

www.elysium.ltd.uk ***(Richard Clark JPEG webmaster*** richard@elysium.ltd.uk)

HD Photo

HDPhoto_v10.doc

(JPEG Convener Daniel Lee
daniel.t.lee@ebay.com convener@jpeg.org)

Microsoft HD Photo Specification

http://www.microsoft.com/whdc/xps/wmphotoeula.mspx

JXR-P1. The authors in [JXR7] state, "that they have presented lifting structures for hierarchical lapped transforms (HLTs) highly compatible with the JPEG XR standard [JXR1] and are lower in complexity in terms of the number of operations and lifting steps compared with JPEG XR. These lifting structures will have any signal processing and communication applications because they are based on commonly used rotation matrices". Explore these applications and see how these lifting structures for HLTs can be used.

JPEG-LS

JLS1. M. J. Weinberger, G. Seroussi and G. Sapiro, "The LOCO-I Lossless Image Compression Algorithm: Principles and Standardization into JPEG-LS," IEEE Trans. on Image Processing, vol. 9, pp. 1309–1324, Aug. 2000.
Website: http://www.hpl.hp.com/loco/HPL-98-193RI.pdf

JLS2. J. Weinberger, G. Seroussi, and G. Sapiro, "LOCO-I: A low complexity, context-based, lossless image compression algorithm", Hewlett-Packard Laboratories, Palo Alto, CA.

JLS3. Ibid, "LOCO-I A low Complexity Context-based, lossless image compression algorithm", Proc. 1996 DCC, pp. 140–149, Snowbird, Utah, Mar. 1996.

JLS4. Z. Zhang, R Veerla and K.R. Rao, "A modified advanced image coding" CANS University Press, pp. 110–116, 2010.

JLS5. UBC JPEG-LS codec implementation [online]. Available: http://www.stat.columbia.edu/~jakulin/jpeg-ls/mirror.htm

JLS6. http://www.ece.ubc.ca/spmg/research/jpeg/jpeg_ls/jpegls.html

JLS7. LOCO-I Software (public domain) (JPEG-LS)
http://spmg.ece.ubc.ca/research/jpeg/jpeg_ls/jpegls.html

JLS8. LOCO_I technical report www.hpl.hp.com/loco (HP Labs)

JLS9. JPEG-LS (LOCO-I) http://www.hpl.hp.com/loco/HPL-98-193R1.pdf

Project

JLS-P1. [JLS4] involves comparison of various image coding standards such as JPEG, JPEG-LS, JPEG2000, JPEG XR, advanced image coding (AIC) and modified AIC (MAIC) (Also H.264/MPEG4 AVC intra mode only). Extend this comparison to HEVC intra mode only. Consider test images at various spatial resolutions and at different bit rates. Include BD-bit rate and BD-PSNR as metrics besides PSNR and SSIM. Consider also implementation complexity.

JPEG

JPEG1. G. K. Wallace, "The JPEG still picture compression standard," Commun. ACM, vol. 34, no. 4, pp. 30–44, Apr. 1991. Also "The JPEG still picture compression standard," IEEE Trans. CE, vol. 38, pp. 18–34, Feb. 1992.

JPEG2. Independent JPEG Group. [Online]. Available: http://www.ijg.org, www.jpeg.org

JPEG3. J. Aas, Mozilla Advances JPEG Encoding With Mozjpeg 2.0. [Online]. Available: https://blog.mozilla.org/research/2014/07/15/mozilla-advances-jpeg-encoding-with-mozjpeg-2-0/
accessed July 2014.

JPEG4. W. B. Pennebaker and J. L. Mitchell, "JPEG Still Image Data Compression Standard," Van Nostrand Reinhold, New York, 1992.

JPEG5. JPEG reference software Website: ftp://ftp.simtel.net/pub/simtelnet/msdos/graphics/jpegsr6.zip

JPEG6. F. Huang et al., "Reversible data hiding in JPEG images", IEEE Trans. CSVT, vol. 26, pp. 1610–1621, Sept. 2016. Several papers related to reversible and other data hiding schemes are listed as references at the end.

JPEG7. X. Zhang et al., "Lossless and reversible data hiding in encrypted images with public key cryptography", IEEE Trans. CSVT, vol. 26, pp. 1622–1631, Sept. 2016.

JPEG8. P.A.M. Oliveira at al, "Low-complexity image and video coding based on an approximate discrete Tchebichef transform", IEEE Trans. CSVT (early access).

JPEG9. T. Richter, "JPEG on steroids: Common optimization techniques for JPEG image compression", IEEE ICIP2016, Phoenix, AZ, Sept. 2016.

JPEG10. H. Lee et al., "A novel scheme for extracting a JPEG image from an HEVC compressed data set" IEEE ICCE, Las Vegas, Jan. 2017.

JPEG11. C. Liu et al, "Random walk graph Laplacian-based smoothness prior for soft decoding of JPEG images", IEEE Trans. IP, vol. 26, pp. 509–524, Feb. 2017.

JPEG12. M.M. Siddeq and M.A. Rodrigues, "A novel high-frequency encoding algorithm for image compression", EURASIP J. on advances in signal processing", 2017:2017:26.

BOSSbase image data set http://www.agents.cz/boss/BOSSFinal IMAGE DATA SET

JPEG-P1 Oliveira et al. [JPEG8] have developed a new low complexity approximation the discrete Tchebichef transform (DTT) and embedded it in JPEG and H.64/AVC. Extend this to 16-point DTT and develop a fast algorithm. Draw a flow graph similar to that shown in Figure 4.

JPEG-P2 Develop hardware aspects of the 16-point DTT as described in VI Hardware section. This implies that the 16-point DTT needs to be implemented on Xilinx FPGA board.

JPEG-P3 See JPEG-P1. Embed the new DTT in HEVC Main profile and compare its performance with anchor HEVC. (See Figure 4).

JPEG-P4 Huang et al. [JPEG6] have proposed a new histogram shifting based RDH (reversible data hiding) scheme in JPEG images that realize high embedding capacity and good visual quality while preserving the JPEG file size. Implement this scheme and confirm the results as described in the figures.

JPEG-P5 See JPEG-P4. In section IV Conclusions the authors state that the proposed novel block selection strategy can result in better visual quality and less JPEG file size. They also state that this technique may be applied to other RDH schemes to improve their performance. Several references related to RDH in images are listed at the end in [JPEG6]. Apply this strategy in the RDH schemes and evaluate their performances. Use different test images at various quality factors. (See also the references listed at the end in [JPEG7]).

JPEG-P6 See [JPEG10] Implement JPEG extraction from the HEVC bit stream in the compressed domain. Confirm the results shown in Table I for the test sequences.

JPEG-P7 See [JPEG11]. In conclusion the authors state "In future work, we will work on speeding up the proposed algorithm to make it more practical". Explore this.

JPEG XT

JXT1. T. Richter et al., "The JPEG XT suite of standards: status and future plans," [9599-30], SPIE. Optics + Photonics, San Diego, California, USA, 9–13, Aug. 2015.

JXT2. M. Fairchild : "The HDR Photographic survey," available online at http://rit-mcsl.org/fairchild/HDR.html (retrieved July 2015).

JXT3. R. Mantiuk, "pfstools : High Dynamic Range Images and Video," available online at http://pfstools.sourceforge.net/ (retrieved July 2015).

JXT4. T. Ricther, "JPEG XT Reference Codec 1.31 (ISO License)," available online at http://www.jpeg.org/jpegxt/software.html (retrieved July 2015).

JXT5. T. Richter, "Lossless Coding Extensions for JPEG," IEEE Data Compression Conference, pp. 143–152, Mar. 2015.

See [JXT1] JPEG XT is a standardization effort targeting the extension of the JPEG features by enabling support for high dynamic range imaging, lossless and near lossless coding and alpha channel coding, while also guaranteeing backward and forward compatibility with the JPEG legacy format. JPEG XT has nine parts described in detail in Figure 2. Further extensions relate to JPEG Privacy and Security and others such as JP Search and JPEG systems and are listed in the conclusion. JPEG XT is forward and backward compatible with legacy JPEG unlike JPEG-LS and JPEG2000.

JXT6 A. Artusi et al., "Overview and Evaluation of the JPEG XT HDR Image Compression Standard", J. of real time image processing, vol. 10, Dec. 2015.

For ready reference, the abstract is reproduced below.

Abstract: Standards play an important role in providing a common set of specifications and allowing interoperability between devices and systems. Until recently, no standard for High Dynamic Range (HDR) image coding had been adopted by the market, and HDR imaging relies on proprietary and vendor specific formats which are unsuitable for storage or exchange of such images. To resolve this situation, the JPEG Committee is developing a new coding standard called JPEG XT that is backwards compatible to the popular JPEG compression, allowing it to be implemented using standard 8-bit JPEG coding hardware or software. In this paper, we present design principles and technical details of JPEG XT. It is based on a two-layers design, a base layer containing a Low Dynamic Range (LDR) image accessible to legacy implementations, and an extension layer providing the full dynamic range. The paper introduces three of currently defined profiles in JPEG XT, each constraining the common decoder architecture to a subset of allowable configurations. We assess the coding efficiency of each profile extensively through subjective assessments, using 24 naive subjects to evaluate 20 images, and objective evaluations, using 106 images with five different tone-mapping operators and at 100 different bit rates. The objective results (based on benchmarking with subjective scores) demonstrate that JPEG XT can encode HDR images at bit rates varying from 1:1 to 1:9 bit/pixel for estimated mean opinion score (MOS) values above 4:5 out of 5, which is considered as fully transparent in many applications. This corresponds to 23-times bit stream reduction compared to lossless OpenEXR PIZ compression.

In the conclusions the authors [JXT6] state "The benchmarking showed that, in terms of predicting quality loss due to coding artifacts, simple metrics, such as PSNR, SNR, and MRSE computed in linear space are unsuitable for measuring perceptual quality of images compressed with JPEG XT, however, the prediction of these metrics improve when applied to the pixels converted in the perceptually uniform space. Also, HDR-VDP-2 provides the best performance as compared to other tested metrics." This again confirms that that these objective metrics do not correlate with the subjective (perceptual) quality.

JXT7 M. Shah, "Future of JPEG XT: Scrambling to enhance privacy and security", M.S. Thesis, EE Dept., UTA, Arlington, TX, USA. http://www.uta.edu/faculty/krrao/dip click on courses and then click on EE5359 Scroll down and go to recent Theses/Project Title and click on Maitri Shah.

JXT8 R. K. Mantiuk, T. Richter and A. Artusi, "Fine-tuning JPEG-XT compression performance using large-scale objective quality testing", IEEE ICIP, Phoenix, Arizona, Sept. 2016.
Two publicly available (image datasets) used in [JXT6]
Fairchild's HDR photographic survey.
http://rit-mcsl.org/fairchild/HDR.html
HDR-eye dataset of HDR images.
http://mmspg.ep.ch/hdr-eye
The dataset contained scenes with architecture, landscapes, portraits, frames extracted from HDR video, as well as computer generated images.

JXT9 P. Korshunov et al., "EPFL's dataset of HDR images" 2015 (online), Available: http://mmsp.epfl.ch/hdr-eye

JXT10 T. Richter, "JPEG on steroids: Common optimization techniques for JPEG image compression", IEEE ICIP, Phoenix, AZ, Sept. 2016.

JPEG Software

JXT11 J. Aas: "Mozilla JPEG encoder project", available online at https://github.com/mozilla/mozjpeg (See [JXT10]).

JXT12 T. Richter, "Error bounds for HDR image coding with JPEGXT", IEEE DCC, April 2017.

Mozilla JPEG Encoder Project

This project's goal is to reduce the size of JPEG files without reducing quality or compatibility with the vast majority of the world's deployed decoders.

The idea is to reduce transfer times for JPEGs on the Web, thus reducing page load times.

'mozjpeg' is not intended to be a general JPEG library replacement. It makes tradeoffs that are intended to benefit Web use cases and focuses solely on improving encoding. It is best used as part of a Web encoding workflow. For a general JPEG library (e.g. your system libjpeg), especially if you care about decoding, we recommend libjpeg-turb libjpeg.

A complete implementation of 10918-1 (JPEG) coming from jpeg.org (the ISO group) with extensions for HDR currently discussed for standardization.

JXT13 Jpegxt demo software T. Richter: "libjpeg: A complete implementation of 10918-1 (JPEG)", available online at https://github.com/thorfdbg/libjpeg (See [JXT10]).

JPEG XT Projects

The JPEG XT suite of standards described in the paper [JXT1] can lead to several projects. Review this paper in detail and implement these standards. Please access the reference software.

JXT-P1. JPEG Privacy & Security: With social media having a huge impact on every individual, it disrupts their privacy. Hence, protecting the privacy and security is becoming very important, not only because of the social media, but also because images/meta-data use cloud for storage purposes over private repositories. Encrypting the images and providing access to only the authorized person can help in maintaining the privacy of the image/meta-data.

JXT-P2. Recoding Image Editing Operations: While editing any image or text, only the original data and the final output after editing is usually saved. But instead, if every intermediate step is also recorded, it gives a possibility to revert the data to the original image or any of the intermediate steps. Saving every step while developing a special effect (sharpen, emboss, smooth) can help to develop new, faster and easier effects.

JXT-P3. In the conclusion section the authors [JXT6] state
"The encoding or decoding hardware can be in fact designed based on a pair of existing JPEG coding chips, as shown in Figure 2, resulting in a minimal hardware change in the existing hardware infrastructure without influencing its real-time performances." Implement the encoding and decoding hardware.

JXT-P4. See [JXT7] As part of JPEG XT future extensions, JPEG privacy and security, Shah has developed an algorithm to encrypt and decrypt the images using C++ and OpenCV libraries. This algorithm performs well for protecting an individual's identity by blurring the image. By resolving the compatibility issues between the OpenCV libraries and JPEG XT code this algorithm can be incorporated in the JPEG XT code to provide the security feature automatically. Explore this.

JXT-P5 See JXT-P4. Shah suggests that the amount of blurring can be changed according to the user preference by changing the kernel size. This gives the user full control over how the image should be seen by others. Explore this in detail and draw some conclusions.

JXT-P6 See JXT-P4. Several other tasks also are proposed as future work. Investigate and implement these tasks.

JXT-P7 See [JXT10], [JXT11] and [JXT12]. Richter has proposed some optimization techniques to improve the legacy JPEG in terms of rate distortion performance close to JPEG-XR (See Figure 5 in [JXT10]). Using the software described in [JXT 11] and [JXT12] confirm the enhanced JPEG rate distortion performance shown in [JXT10].

JXT-P8 JPEG-XT standard defines a normative decoding procedure to reconstruct an HDR image from two JPEG regular code streams named the base layer (visible to legacy decoders) and an extension layer and how to merge them together to form one single image. See also [JXT1]. The standard does not however, define the encoding procedure and leaves large freedoms to the encoder for defining the necessary decoder configuration. Mantiuk, Richter and Artusi [JXT8], explored the whole space of possible configurations to achieve the best possible R-D performance. Review this paper in detail and verify the performance results described in the figures [JXT8].

JPEG PLENO

JP1. T. Ebrahimi, "Towards a new standard for plenoptic image compression", IEEE DCC, 29 March – 1 April 2016. Abstract of this valuable paper is reproduced below:

Abstract

JPEG format is today a synonymous of modern digital imaging, and one of the most popular and widely used standards in recent history. Images created in JPEG format now exceed one billion per day in their number, and most of us can count a couple, if not more JPEG codecs in devices we regularly use in our daily lives; in our mobile phones, in our computers, in our tablets, and of course in our cameras. JPEG ecosystem is strong and continues an exponential growth for the foreseeable future. A significant number of small and large successful companies created in the last two decades have been relying on JPEG format, and this trend will likely continue.

A question to ask ourselves is: will we continue to have the same relationship to flat snapshots in time (the so-called Kodak moments) we call pictures, or could there be a different and enhanced experience created when capturing and using images and video, that could go beyond the experience images have been providing us for the last 120 years? Several researchers, artists, professionals, and entrepreneurs have been asking this same question and attempting to find answers, with more or less success. Stereoscopic and multi-view photography, panoramic and 360-degree imaging, image fusion, point cloud, high dynamic range imaging, integral imaging, light field imaging, and holographic imaging are among examples of solutions that have been proposed as future of imaging.

Recent progress in advanced visual sensing has made it feasible to capture visual content in richer modalities when compared to conventional image and video. Examples include Kinect by Microsoft, mobile sensors in Project Tango by Google and Intel, light-field image capture by Lytro, light-field video by Raytrix, and point cloud acquisition by LIDAR (Light Detection and Ranging). Likewise, image and video rendering solutions are increasingly relying on richer modalities offered by such new sensors. Examples include Head Mounted Displays by Oculus and Sony, 3D projector by Ostendo and 3D light field display solutions by Holografika. This promises a major change in the way visual information is captured, processed, stored, delivered and displayed.

JPEG PLENO evolves around an approach called plenoptic representation, relying on a solid mathematical concept known as plenoptic function. This promises radically new ways of representing visual information when compared to traditional image and video, offering richer and more holistic information. The plenoptic function describes the structure of the light information impinging on observers' eyes, directly measuring various underlying visual properties like light ray direction, multi-channel colors, etc.

The road-map for JPEG PLENO follows a path that started in 2015 and will continue beyond 2020, with the objective of making the same type of impact that the original JPEG format has had on today's digital imaging starting from 20 years ago. Several milestones are in work to approach the ultimate image representation in well-thought, precise, and useful steps. Each step could potentially offer an enhanced experience when compared to the previous, immediately ready to be used in applications, with potentially backward compatibility. Backward compatibility could be either at the coding or at the file format level, allowing an old JPEG decoder of 20 years ago to still be able to decode an image, even if that image won't take full advantage of the intended experience, which will be only offered with a JPEG PLENO decoder.

This talk starts by providing various illustrations the example applications that can be enabled when extending conventional image and video models toward plenoptic representation. Doing so, we will discuss use cases and application requirements, as well as example of potential solutions that are or could be considered to fulfill them. We will then discuss the current status of development of JPEG PLENO standard and discuss various milestones ahead. The talk will conclude with a list of technical challenges and other considerations that need to be overcome for a successful completion of JPEG PLENO.

JP2. M.D. Angelo et al., "Correlation plenoptic imaging", Physical Review Letters, vol. 116, no. 22, 602, Jun. 2016.

Abstract is reproduced below.

Plenoptic imaging is a promising optical modality that simultaneously captures the location and the propagation direction of light in order to enable tridimensional imaging in a single shot. However, in classical imaging systems, the maximum spatial and angular resolutions are fundamentally linked; thereby, the maximum achievable depth of field is inversely proportional to the spatial resolution. We propose to take advantage of the second-order correlation properties of light to overcome this fundamental limitation. In this paper, we demonstrate that the momentum/position correlation of chaotic light leads to the enhanced refocusing power of correlation plenoptic imaging with respect to standard plenoptic imaging.

JP-P1 Please review the abstract in [JP1]. Focus on the research towards developing the JPEG PLENO standard with backward compatibility with legacy JPEG. Review the technical challenges and consider the application requirements. This is a challenging research.

JP-P2 Please review [JP2] in detail. In the conclusions and outlook, the authors state that the merging of plenoptic imaging and correlation quantum imaging has thus the potential to open a totally new line of research. Explore this.

JPEG XS

The overview paper on HEVC SCC extension [OP11] states that the JPEG has initiated the standardization of JPEG XS [JXS1, JXS3] a low latency lightweight image coding system with potential applications in video link compression, frame buffer compression, and real time video storage. This paper compares the requirements and features of HEVC SCC extension, DSC and JPEG XS in Table V with the foot note that the requirements and features of JPEG XS are subject to change. The JPEG XS international standard is projected to be finalized in mid 2018. Willeme et al. [JXS2] conducted preliminary comparison of several existing lightweight compression schemes.

JXS1 JPEG Committee Initiates Standardization of JPEG XS a Low-Latency Lightweight Image Coding System, July 2015 [online] Available: http://jpeg.org/items/20150709_press.html.

JXS2 A. Willeme, et al., "Quality and error robustness assessment of low-latency lightweight intra-frame codecs" Proc. IEEE Data Compression Conf. (DCC), Mar. 2016.

JXS3 JPEG XS—Call for Proposals for a Low-Latency Lightweight Image Coding System, Mar. 2016 [online] Available: https://jpeg.org/items/20160311_cfp_xs.html.

JXS-P1 Review the web sites on JPEG XS [JXS1] and [JXS2] and submit proposals to JPEG that can meet the requirements of JPEG XS. Note that this project requires group effort.

LAR-LLC

LL1. Y. Liu, O. Deforges and K. Samrouth, "LAR-LLC: A low complexity multiresolution lossless image codec," IEEE Trans. CSVT, vol. 26, pp. 1490–1501, Aug 2016.

Abstract

It compares the performance of LAR-LLC lossless codec with JPEGXR, JPEG2000, JPEGLS and LJPEG (lossless JPEG) in terms of compression ratio, encoding and decoding speeds (see Tables I–IV). LAR-LLC has much less encoding and decoding speeds (less complexity) compared to the JPEG series. It also supports spatial scalability similar to JPEG2000.

LL-P1. See [LL1]. In Table II the bitrates (bpp) for lossless color image (24 bpp) codecs are shown, for various test images. Implement the JPEG lossless codecs and compare with LAR-LLC.

LL-P2. HEVC has lossless coding in all intra (AI) profile. Implement this and compare with Table II.

LL-P3. Develop Tables similar to Tables V and VI for HEVC AI profile (lossless coding).

LL-P4. In the paragraph before VII conclusion, the authors state that the LAR-LLC currently runs in a mono thread configuration, but has potential parallel processing ability in multithreading and suggest how this can be implemented. Explore this and evaluate how the multithreading configuration can further reduce the LAR-LLC encoding/decoding speeds for lossless coding.

LL-P5. Extend the lossless image coding scheme to VP10. See Tables I to IV in [LL1].

LL-P6. See LL-P4. Extend the lossless image coding scheme to AVS2 – intra coding only.

PNG

PNG1. T. Boutell, "PNG (Portable Network Graphics) Specification version 1.0", Network Working group RFC 2083, available online at http.ietf.org/html/rfc2083 (retrieved November 2014).

WebP

Web.P1. *WebP—Google Developers*. [Online]. Available: http://code.google.com/speed/webp/

Web.P2. The WebP container format is specified at https://developers.google.com/speed/webp/docs/riff_container, while the VP8L format is specified at https:// developers.google.com/speed/webp/docs/webp_lossless_bitstream_specification.

WebM

WebM1. The WebM Project. [Online]. Available: http://www.webmproject.org/

WebM2. WebM™: an open web media project, VP8 Encode Parameter Guide, 2-Pass Best Quality VBR Encoding, Online: http://www.webmproject.org/docs/encoder-parameters/#2-pass-best-qualityvbr-encoding

WebM3. Paul Wilkins, Google® Groups "WebM Discussion", Online: https://groups.google.com/a/webmproject.org/forum/?fromgroups#!topic/webmdiscuss/UzoX7owhwB0

WebM4. Ronald Bultje, Google® Groups "WebM Discussion", Online: https://groups.google.com/a/webmproject.org/forum/?fromgroups#!topic/webmdiscuss/xopTll6KqGI

WebM5. John Koleszar, Google® Groups "Codec Developers", Online: https://groups.google.com/a/webmproject.org/forum/#!msg/codecdevel/yMLXzaohONU/m69TbYnEamQJ

DIRAC (BBC)

[DR1] T. Borer and T. Davies, "Dirac – video compression using open technology," *EBU Technical Review*, pp. 1–9, July 2005.

[DR2] Dirac website: www.diracvideo.org

[DR3] Dirac software: http://sourceforge.net/projects/dirac/

[DR4] A. Urs, "Multiplexing/demultiplexing Dirac video and AAC audio while maintaining lip sync", M.S. Thesis, EE Dept., University of Texas at Arlington, Arlington, Texas, May 2011.

[DR5] A. Ravi and K.R. Rao, "Performance analysis and comparison of the Dirac video codec with H.264/MPEG-4, Part 10", Intrnl. J. of wavelets, multi resolution and information processing (IJWMIP), vol. 4, No. 4, pp. 635–654, 2011.

[DR6] See DR5. Implement performance analysis and comparison of the Dirac video codec with HEVC.

[DR7] BBC Research on Dirac: http://www.bbc.co.uk/rd/projects/dirac/index.shtml

[DR8] A. Mandalapu, "STUDY AND PERFORMANCE ANALYSIS OF HEVC, H.264/AVC AND DIRAC", project report, Spring 2016. http://www.uta.edu/faculty/krrao/dip click on courses and then click on EE5359 Scroll down and go to projects Spring 2016 and click on A. Mandalapu final report.

[DR9] S.S.K.K. Avasarala, "Objective Video quality assessment of Dirac and H.265" project report, Spring 2016.

http://www.uta.edu/faculty/krrao/dip click on courses and then click on EE5359 Scroll down and go to projects Spring 2016 and click on S.S.K.K. Avasarala final report.

[DR10] http://diracvideo.org/2012/01/schroedinger-1-0-11/ access to DIRAC reference software.

[DR11] https://en.wikipedia.org/wiki/Dirac (video compression format) – website on DIRAC.

[DR12] https://www.youtube.com/watch?v=HVFFq44UvLA A detailed tutorial about dirac video codec by BBC.

DR.P1 See [DR8] Performance comparison of HEVC, DIRAC and H.264/AVC using CIF and QCIF test sequences shows, in general, superior performance of HEVC. Extend this comparison to SD, HDTV, UHDTV, 4K and 8K test sequences. See also [DR9].

DR.P2 Fracastoro, Fossen and Magli [E392] developed steerable DCT (SDCT) and its integer approximation and applied them to image coding. They also compared this with wavelet coding (Table III). Confirm the results shown in this Table and extend this to higher resolution images.

DR.P3 See [E393]. Apply the FCDR to DIRAC codec both as a post processing operation and as an in loop (embedded) operation. Implement the projects for DIRAC similar to those described in P.5.268 and P.5.269.

DAALA

DAALA1 T.J. Daede et al., "A perceptually-driven next generation video codec", IEEE DCC, Snow Bird, Utah, March–April 2016. Abstract of this paper is repeated here:

The Daala project is a royalty-free video codec that attempts to compete with the best patent-encumbered codecs. Part of our strategy is to replace core tools of traditional video codecs with alternative approaches, many of them designed to take perceptual aspects into account, rather than optimizing for simple metrics like PSNR. This paper documents some of our experiences with these tools, which ones worked and which did not, and what we have learned from them. The result is a codec which compares favorably with HEVC on still images, and is on a path to do so for video as well.

DAALA2 "Daala test media", https://wiki.xiph.org/Daala Quickstart#Test Media.

https://xiph.org/daala
https://git.xiph.org/?p=daala.git
Daala Codec demo –
http://people.xiph.org/∼xiphmont/demo/daala/demo1.shtml

Daala Video Compression

Daala is the code-name for a new video compression technology. The effort is a collaboration between Mozilla Foundation, Xiph.Org Foundation and other contributors.

The goal of the project is to provide a free to implement, use and distribute digital media format and reference implementation with technical performance superior to h.265.

Technology Demos

- Next generation video: Introducing Daala
- Introducing Daala part 2: Frequency Domain Intra Prediction
- Introducing Daala part 3: Time/Frequency Resolution Switching
- Introducing Daala part 4: Chroma from Luma
- Daala: Painting Images For Fun (and Profit?)
- Daala: Perceptual Vector Quantization (PVQ)

DAALA3 R.R. Etikala, "Performance comparison of DAALA and HEVC", project report, Spring 2016. http://www.uta.edu/faculty/krrao/dip click on courses and then click on EE5359 Scroll down and go to projects Spring 2016 and click on R.R. Etikala final report. This report implemented both HEVC and DALLA and compares their performances in terms of R-D plots for CIF and QCIF test sequences.

DAALA4 J.-M. Valin et al., "DAALA: A perceptually-driven still picture codec", IEEE ICIP, Session MA.L3: Image compression grand challenge, Phoenix, AZ, Sept. 2016.

DAALA-P1 Implement and improve Daala video codec using video test sequences and compare with HEVC/H.264/VP9 and other standard based codecs using implementation complexity, PSNR, SSIM, BD-bit rate and BD-PSNR as the metrics. See [DAALA1 and DAALA3].

DAALA-P2 Performance comparison of SDCT with DCT and DDCT is investigated in [E392]. In DAALA INTDCT is used. Replace INTDCT with INTSDCT and implement the performance comparison based on various test sequences including 4K and 8K and different block sizes. See the Tables and Figures in [E392].

DAALA-P3 See [E393]. Apply the FCDR to DAALA video codec both as a post processing operation and as an in loop (embedded) operation. Implement the projects for DAALA similar to those described in P.5.268 and P.5.269.

MPEG-DASH

MD1. I. Irondi, Q. wang and C. Grecos, "Subjective evaluation of H.265/HEVC based dynamic adaptive video streaming over HTTP (HEVC-DASH), SPIE-IS&T Electronic Imaging, vol. 9400, p. 94000B (session: real-time image and video processing), 2015.

MD.P1 See [MD1]. Irondi, Wang and Grecos have presented a subjective evaluation of a HEVC-DASH system that has been implemented on a hardware test bed. In the abstract the authors state "DASH standard is becoming increasingly popular for real-time adaptive HTTP streaming of internet video in response to unstable network conditions. Integration of DASH streaming technologies with the new H.265/HEVC video coding standard is a promising area of research." In the conclusions, they state "The results of this study can be combined with objective metrics to shed more light on the (QOE) – quality of experience – of DASH systems and correlation between the QOE results and the objective performance metrics will help designers in optimizing the performance of DASH systems". Papers listed as references in [MD1] relate to both subjective and objective performances (each separately) of DASH systems. Combine (integrate) both objective and subjective criteria with HEVC-DASH and evaluate the QOE. Based on this evaluation optimize the performance of HEVC-DASH system.

Bjontegaard Metric

http://www.mathworks.com/matlabcentral/fileexchange/27798-bjontegaard metric/content/bjontegaard.m

BJONTEGAARD metric calculation: Bjontegaard's metric allows to compute the average gain in PSNR or the average per cent saving in bitrate between two rate-distortions curves. See [E81], [E82], [E96] and [E198]. This metric reflects the subjective visual quality better than MSE and PSNR.

Figure below explains clearly the BD-bit rate and BD-PSNR [E202].

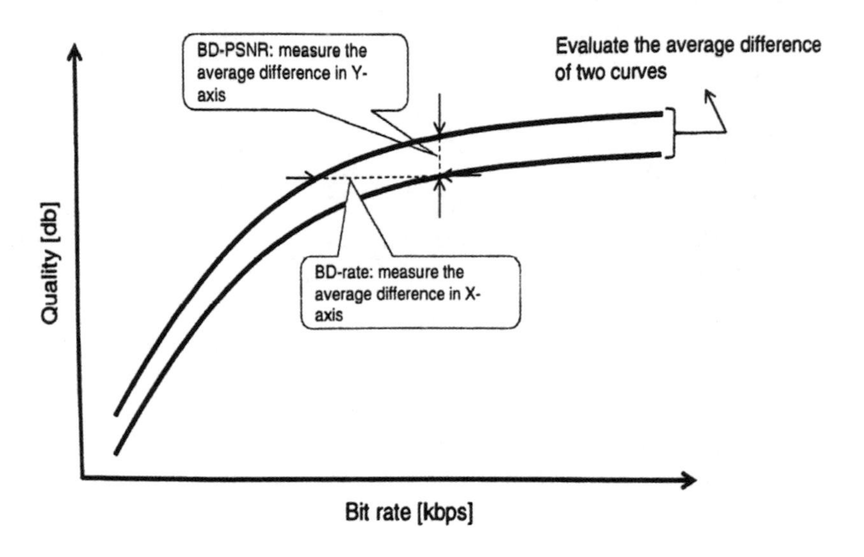

AVS China

[AVS1] AVS Video Expert Group, "Information Technology – Advanced coding of audio and video – Part 2: Video (AVS1-P2 JQP FCD 1.0)," Audio Video Coding Standard Group of China (AVS), Doc. AVS-N1538, Sep. 2008.

[AVS2] AVS Video Expert Group, "Information technology – Advanced coding of audio and video – Part 3: Audio," Audio Video Coding Standard Group of China (AVS), Doc. AVS-N1551, Sep. 2008.

[AVS3] L. Yu *et al.*, "Overview of AVS-Video: Tools, performance and complexity," *SPIE VCIP*, vol. 5960, pp. 596021-1–596021-12, Beijing, China, July 2005.

[AVS4] L. Fan, S. Ma and F. Wu, "Overview of AVS video standard," *IEEE Int'l Conf. on Multimedia and Expo, ICME '04*, vol. 1, pp. 423–426, Taipei, Taiwan, June 2004.

[AVS5] W. Gao et al., "AVS – The Chinese next-generation video coding standard," National Association of Broadcasters, Las Vegas, 2004.

[AVS6] AVS-China official website: http://www.avs.org.cn

[AVS7] AVS-china software download: ftp://159.226.42.57/public/avs_doc/avs_software

[AVS8] IEEE Standards Activities Board, "AVS China - Draft Standard for Advanced Audio and Video Coding", adopted by IEEE as standard IEEE 1857. Website: stds.ipr@ieee.org

[AVS8] S. Ma, S. Wang and W. Gao, "Overview of IEEE 1857 video coding standard," IEEE ICIP, pp. 1500–1504, Sept. 2013.

[AVS9] D. Sahana and K.R. Rao, "A study on AVS-M standard", Calin ENACHESCU, Florin Gheorghe FILIP, Barna Iantovics (Eds.), Advanced Computational Technologies published by the Romanian Academy Publishing House, pp. 311–322, Bucharest, Rumania, 2012.

[AVS10] S. Sridhar, "Multiplexing/De-multiplexing AVS video and AAC audio while maintaining lip sync", M.S. Thesis, EE Dept., University of Texas at Arlington, Arlington, Texas, Dec. 2010. http://www.uta.edu/faculty/krrao/dip click on courses and then click on EE5359 Scroll down and go to Thesis/Project Title and click on S. Sridhar.

[AVS11] W. Gao and S. Ma, "Advanced video coding systems", Springer, 2015.
Audio and video standard China adopted as IEEE 1857.4

[AVS12] K. Fan et al., "iAVS2: A fast intra encoding platform for IEEE 1857.4", IEEE Trans. CSVT, early access. Several papers related to fast intra coding in HEVC are listed at the end.

[AVS13] Z. He, L. Yu, X. Zheng, S. Ma and Y. He, "Framework of AVS2 Video Coding", IEEE Int. Conf. Image Process., pp. 1515–1519. Sept. 2013.

[AVS14] S. Ma, T. Huang and C. Reader, "AVS2 making video coding smarter [standards in a nutshell]", IEEE Signal Process. Mag., vol. 32, no. 2, pp. 172–183, 2015.

[AVS15] X. Zhang et al., "Adaptive loop filter for avs2", in Proc. 48th AVS Meeting, 2014.

[AVS16] Screen and mixed content coding working draft 1, Document AVS N2283, Apr. 2016.

[AVS17] Common test conditions for screen and mixed content coding, Document AVS N2282, Apr. 2016.

AVS-P1 Review the references listed above on AVS China. In particular review [AVS8]. Implement various profiles in AVS China and compare its performance with HEVC using various video test sequences. Comparison needs to be based on various metrics such as PSNR, SSIM, BD-bitrate and BD-PSNR besides computational complexity. The profiles involve several projects.

AVS-P2 See [AVS 12]–[AVS 15] AVS2 is the second generation of the audio video coding standard to be issued as IEEE 1857.4 standard. It doubles the coding efficiency of AVS1 and H.264/AVC. Implement this standard and compare with HEVC based on standard performance metrics using various test sequences at different spatial/temporal resolutions.

AVS-P3 See AVS-P2. In the conclusions, the authors [AVS 12] state "Owing to their similar frameworks, the proposed systematic solution and the fast algorithms can also be applied in HEVC intra encoder design". Explore this.

AVS-P4 Fan et al. [AVS12] proposed the fast intra encoding platform for AVS2 using numerous speedup methods. Using the test sequences shown in Table IV confirm the results shown in Tables V–X and Figures 13–16.

AVS-P5 Compare the performance of AI image coding techniques including JPEG series, HEVC, MPEG-4/AVC, DIRAC, DAALA, THOR, AVS1, AVS2, VP8-10 and AV1 (developed by Alliance for Open Media – AOM). As always use various test sequences at different spatial resolutions.

AVS-P6 Performance comparison of SDCT with DCT and DDCT is investigated in [E392]. In AVS1 and AVS2 INTDCT is used. Replace INTDCT with INTSDCT and implement the performance comparison based on various test sequences including 4K and 8K and different block sizes. See the Tables and Figures in [E392].

AVS-P7 See [E393]. Apply the FCDR to AVS China codec (AVS2) both as a post processing operation and as an in loop (embedded) operation.

Implement the projects for AVS2 similar to those described in P.5.268 and P.5.269.

AVS-P8 The AVS Workgroup of China has released the first working draft [AVS16] of the AVS SCC extension along with the reference software and common test conditions [AVS17] using the test sequences [TS27] developed for SCC. Implement the AVS China SCC extension and compare its performance with the HEVC SCC extension using the standard metrics. See [OP11]. There may be a restricted access to both [AVS16] and [AVS17].

THOR Video Codec

Website: http://tinyurl.com/ismf77

TVC1 G. Bjontegaard et al., " The Thor video codec", IEEE DCC, pp. 476-485, Sun Bird, Utah, March–April 2016.
Thor video codec is being developed by Cisco Systems as a royalty free codec. Details regarding block structure, intra/inter prediction, fractional sample interpolation, motion vector coding, transform, quantization, in-loop filtering, entropy coding, transform coefficient scanning and other functionalities are described in [TVC1]. Encoder follows the familiar motion compensated, transform predictive coding as in AVC/H.264 and HEVC/H.265. For performance comparison of TVC with VP9 and x265 see Table 3 (low delay and high delay) and Figures 5 and 6. For this comparison only HD sequences (classes B and E) and 1080p video conferencing sequences were used.

[TVC-P1] Extend this comparison using all classes of test sequences and develop graphs based on BD-bit rate and BD-PSNR versus encoded bit rate.

[TVC-P2] Consider implementation complexity as another metric and develop tables showing the complexity comparison of TVC with AVC/H.264 and HEVC/H.265 for all test sequences at various bit rates.

[TVC-P3] Carry out extensive research and explore how the various functionalities described in [TVC1] can be modified such that the TVC can be highly competitive with AVC/H.264 and HEVC/H.265. Keep in view that the TVC is royalty free.

[TVC-P4] Performance comparison of SDCT with DCT and DDCT is investigated in [E392]. In TVC, INTDCT is used. Replace

INTDCT with INTSDCT and implement the performance comparison based on various test sequences including 4K and 8K and different block sizes. See the Tables and Figures in [E392].

[TVC-P4] See [E393]. Apply the FCDR to Thor video codec both as a post processing operation and as an in loop (embedded) operation. Implement the projects for TVC similar to those described in P.5.268 and P.5.269.

References on Screen Content Coding

[SCC1] D. Flynn, J. Sole and T. Suzuki, "High efficiency video coding (HEVC) range extensions text specification", Draft 4, JCT-VC. Retrieved 2013-08-07.

[SCC2] M. Budagavi and D.-Y. Kwon, "Intra motion compensation and entropy coding improvements for HEVC screen content coding", IEEE PCS, pp. 365–368, San Jose, CA, Dec. 2013.

[SCC3] M. Naccari et al., "Improving inter prediction in HEVC with residual DPCM for lossless screen content coding", IEEE PCS, pp. 361–364, San Jose, CA, Dec. 2013.

[SCC4] T. Lin et al., "Mixed chroma sampling-rate high efficiency video coding for full-chroma screen content", IEEE Trans. CSVT, vol. 23, pp. 173–185, Jan. 2013.

[SCC5] Braeckman et al., "Visually lossless screen content coding using HEVC base-layer", IEEE VCIP 2013, pp. 1–6, Kuching, China, 17–20, Nov. 2013.

[SCC6] W. Zhu et al., "Screen content coding based on HEVC framework", IEEE Trans. Multimedia, vol. 16, pp. 1316–1326 Aug. 2014 (several papers related to MRC) MRC: mixed raster coding.

[SCC7] ITU-T Q6/16 Visual Coding and ISO/IEC JTC1/SC29/WG11 Coding of Moving Pictures and Audio.

[SCC8] Title: Joint Call for Proposals for Coding of Screen Content Status: Approved by ISO/IEC JTC1/SC29/WG11 and ITU-T SG16 Q6/16 (San Jose, 17 January 2014). Final draft standard expected: late 2015.

[SCC9] F. Zou, et al., "Hash Based Intra String Copy for HEVC Based Screen Content Coding," IEEE ICME (workshop), Torino, Italy, June 2015

[SCC10] T. Lin, et al., "Arbitrary Shape Matching for Screen Content Coding," in Picture Coding Symposium (PCS), pp. 369–372, Dec. 2013.

[SCC11] T. Lin, X. Chen and S. Wang, " Pseudo-2d-matching based dual – coder architecture for screen contents coding," IEEE International Conference in Multimedia and Expo Workshops (ICMEW), pp. 1–4, July 2013.

[SCC12] B. Li, J. Xu and F. Wu, "1d dictionary mode for screen content coding," in Visual Communication and Image Processing Conference, pp. 189–192, Dec. 2014.

[SCC13] Y. Chen and J. Xu, "HEVC Screen Content Coding core experiment 4 (scce4) : String matching for sample coding," JCTVC – Q1124, Apr. 2014.

[SCC14] H. Yu, et al., "Common conditions for screen content coding tests," JCTVC – Q1015, vol. 24, no. 3, pp. 22–25, Mar. 2014.

[SCC15] IEEE Journal on Emerging and Selected Topics in Circuits and Systems (JETCAS)
Special Issue on Screen Content Video Coding and Applications, vol. 6, issue 4, pp. 389–584, Dec. 2016. This has 21 papers on all aspects of screen content video coding including guest editorial, overview paper, future research opportunities and standards developments in this field.
See the overview paper: W.-H. Peng, et al., "Overview of screen content video coding: Technologies, standards and beyond", IEEE JETCAS, vol. 6, issue 4, pp. 393–408, Dec. 2016.

[SCC16] J. Nam, D. Sim and I.V. Bajic, "HEVC-based Adaptive Quantization for Screen Content Videos," IEEE Int. Symp. on Broadband Multimedia Systems, pp. 1–4, Seoul, Korea, 2012.

[SCC17] S.-H. Tsang, Y.-L. Chan and W.-C. Siu, "Fast and Efficient Intra Coding Techniques for Smooth Regions in Screen Content Coding Based on Boundary Prediction Samples", ICASSP2015, Brisbane, Australia, April. 2015.

[SCC18] HEVC SCC Extension Reference software:
https://hevc.hhi.fraunhofer.de/svn/svn_HEVCSoftware/tags/HM-15.0+RExt-8.0+SCM-2.0/

[SCC19] HEVC SCC Software reference manual:
https://hevc.hhi.fraunhofer.de/svn/svn_HEVCSoftware/branches/HM-SCC-extensions/doc/software-manual.pdf

Link to Screen content Test Sequences:
http://pan.baidu.com/share/link?shareid=3128894651&uk=889443731

[SCC20] K. Rapaska et al., "Improved block copy and motion search methods for HEVC screen content coding," [9599-49], SPIE. Optics + photonics, San Diego, California, USA, 9–13, Aug. 2015. Website : www.spie.org/op

[SCC21] B. Li, J. Xu and G. J. Sullivan, "Performance analysis of HEVC and its format range and screen content coding extensions", [9599-45], SPIE. Optics + photonics, San Diego, California, USA, 9–13, Aug. 2015. Website: www.spie.org/op

[SCC22] S. Hu et al., "Screen content coding for HEVC using edge modes", IEEE ICASSP, pp. 1714–1718, 26–31 May 2013.

[SCC23] H. Chen, A. Saxena and F. Fernandes, "Nearest-neighbor intra prediction screen content video coding", IEEE ICIP, pp. 3151–3155, Paris, 27–30, Oct. 2014.

[SCC24] X. Zhang, R. A. Cohen and A. Vetro, "Independent Uniform Prediction mode for screen content video coding," IEEE Visual Communications and Image Processing Conference, pp. 129–132, 7–10 Dec. 2014.

[SCC25] J. Nam, D. Sim and I. V. Bajic, "HEVC-based adaptive quantization for screen content videos," IEEE International Symposium on Broadband Multimedia Systems and Broadcasting (BMSB), pp. 1–4, June 2012.

[SCC26] L. Guo et al., "Color palette for screen content coding", IEEE ICIP, pp. 5556–5560, Oct. 2014.

[SCC27] D.-K. Kwon and M. Budagavi, "Fast intra block copy (intra BC) search for HEVC Screen Content Coding," IEEE ISCAS, pp. 9–12, June 2014.

[SCC28] J. -W. Kang, " Sample selective filter in HEVC intra-prediction for screen content video coding," IET Journals & Magazines, vol. 51, no. 3, pp. 236–237, Feb. 2015.

[SCC29] H. Zhang et al., "HEVC – based adaptive quantization for screen content by detecting low contrast edge regions," IEEE ISCAS, pp. 49–52, May 2013.

[SCC30] Z. Ma et al., "Advanced screen content coding using color table and index map," IEEE Trans. on Image Processing, vol. 23, pp. 4399–4412, Oct. 2014.

[SCC31] R. Joshi et al., "Screen Content Coding Test Model 3 Encoder Description (SCM 3)," Joint Collaborative Team on Video Coding, JCTVC-S1014, Strasbourg, FR, 17–24 Oct. 2014.

[SCC32] S. wang et al., "Joint Chroma Downsampling and Upsampling for screen Content Image", IEEE Trans. on CSVT. (Early Access).

[SCC33] H. yang, Y. Fang and W. Lin, "Perceptual Quality Assessment of Screen Content Images," IEEE Trans. on Image processing. (Early Access).

[SCC34] I. J. S. W. R. Subgroup, "Requirements for an extension of HEVC for coding on screen content," in MPEG 109 meeting, 2014.

[SCC35] H. Yang et al., "Subjective quality assessment of screen conetent images," in International Workshop on Quality of Multimedia Experience (QoMEX), 2014.

[SCC36] "SIQAD," https://sites.google.com/site/subjectiveqa/. : Perceptual Quality Assessment of Screen Content Images.

[SCC37] S. Kodpadi, "Evaluation of coding tools for screen content in High Efficiency Video Coding" M.S. Thesis, EE Dept., University of Texas at Arlington, Arlington, Texas, Dec. 2015. http://www.uta.edu/faculty/krrao/dip click on courses and then click on EE5359 Scroll down and go to recent Theses/Project Title and click on Swetha Kodpadi.

[SCC38] N.N. Mundgemane, "Multi-stage prediction scheme for Screen Content based on HEVC", M.S. Thesis, EE Dept., University of Texas at Arlington, Arlington, Texas, Dec. 2015.
http://www.uta.edu/faculty/krrao/dip click on courses and then click on EE5359 Scroll down and go to recent Theses/Project Title and click on N.N. Mundgemane.

[SCC39] T. Vernier et al., "Guided chroma reconstruction for screen content coding", IEEE Trans. CSVT, (early access).

[SCC40] H. Yu et al., "Requirements for an extension of HEVC for coding of screen content," ISO/IEC JTC1/SC 29/WG 11 Requirements subgroup, San Jose, California, USA, document MPEG2014/N14174, Jan. 2014.

[SCC41] C. Lan et al., "Screen content coding," 2nd JCT-VC meeting, Geneva, Switzerland, document JCTVC-B084, July 2010.

[SCC42] B. Li and J. Xu, "Non-SCCE1: Unification of intra BC and inter modes," 18th JCT-VC meeting, Sapporo, Japan, document JCTVC-R0100, July 2014.

[SCC43] R.R. Barkur, "Production techniques for palette coding in screen content", project report, Spring 2016.
http://www.uta.edu/faculty/krrao/dip click on courses and then click on EE5359 Scroll down and go to projects Spring 2016 and click on R.R. Barkur final report.

[SCC44] C. Chen et al., "A staircase transform coding scheme for screen content video coding" IEEE ICIP, session MPA.P8.1, Phonex, AZ, Sept. 2016.

[SCC45] T. Vermeir et al., "Guided Chroma Reconstruction for Screen Content Coding", IEEE Trans. CSVT, vol. 26, pp. 18184–1892, Oct. 2016.

[SCC46] W. Xiao et al., "Weighted rate-distortion optimization for screen content coding", IEEE Trans. CSVT, (Early access).

[SCC47] C.-C. Chen and W.-H. Peng, "Intra line copy for HEVC screen content coding", IEEE Trans. CSVT, (Early access).

[SCC48] W.-H. Peng et al., "Overview of screen content video coding: Technologies, Standards and Beyond", IEEE JETCAS, vol. 6, issue 4, pp. 393–408, Dec. 2016. This paper is part of the special issue on screen content video coding and applications. IEEE JETCAS, vol. 6, issue 4, pp. 389–584, Dec. 2016. This special issue can lead to several projects on SCC.

Access to JCT-VC Resources

Access to JCT-VC documents: http://phenix.it.-sudparis.eu/jct/
Bug tracking: http://hevc.hhi.fraunhofer.de/trac/hevc
Email list: http://mailman.rwth-aachen.de/mailman/listinfo/jct-vc

Beyond HEVC

[BH1] J. Chen et al., "Coding tools investigation for next generation video coding based on HEVC", [9599-47], SPIE. Optics + photonics, San Diego, California, USA, 9–13, Aug. 2015.

[BH2] A. Alshin et al., "Coding efficiency improvements beyond HEVC with known tools," [9599-48], SPIE. Optics + photonics, San Diego, California, USA, 9–13, Aug. 2015. References [2] through [8] describe various proposals related to NGVC (beyond HEVC) as ISO/IEC JTC – VC documents presented in Warsaw, Poland,

June 2015. Several projects can be implemented/explored based on [BH2] and these references.

[BH3] A. Alexander and A. Elina, "Bi-directional optical flow for future video codec", IEEE DCC, Sun Bird, Utah, March-April 2016.

[BH4] X. Zhao et al., "Enhanced multiple transform for video coding", IEEE DCC, Sun Bird, Utah, March-April 2016.

[BH5] Requirements for a Future Video Coding Standard v1, Standard, ISO/IEC JTC1/SC29/WG11, N15340, June 2015.

[BH6] Algorithm Description of Joint Exploration Test Model 1 (JEM 1), Standard ISO/IEC JTC1/SC29/WG11, N15790, Oct. 2015.

[BH17] S. -H. Park and E.S. Jang, "An efficient motion estimation method for OTBT structure in JVET', IEEE DCC, April 2017.

Projects on Beyond HEVC

[BH-P1] In Alshin et al. [BH2] several tools (some of these are straight forward extensions of those adopted in HEVC) are considered in NGVC (beyond HEVC). Some of these are increasing CU and TU sizes, up to 64 adaptive intra directional predictions, multi – hypothesis probability estimation for CABAC, bi-directional optical flow, secondary transform, rotational transform and multi-parameter intra prediction. Go through [BH2] in detail and evaluate the performance of each new tool, for all intra, random access, low delay B and low delay P (see Table 1). Compare the computational complexity of each tool with that of HEVC.

[BH-P2] See [BH-P1]. Evaluate performance impact of enlarging CU and TU sizes (see Table 2). Also consider computational complexity.

[BH-P3] See [BH-P1]. Evaluate performance impact of fine granularity Intra prediction vs HEVC with enlarged CU and TU sizes (see Table 3). Also consider computational complexity.

[BH-P4] See [BH-P1]. Evaluate performance impact of multi-hypothesis probability estimation vs HEVC with enlarged CU and TU sizes (see Table 4). Also consider computational complexity.

[BH-P.5] See [BH-P1]. Evaluate performance impact of bi-directional optical flow vs HEVC with enlarged CU and TU sizes (see Table 6). Also consider computational complexity.

[BH-P6] See [BH-P1]. Evaluate performance impact of implicit secondary transform vs HEVC with enlarged CU and TU sizes (see Table 7). Also consider computational complexity.

[BH-P7] See [BH-P1]. Evaluate performance impact of explicit secondary transform vs HEVC with enlarged CU and TU sizes (see Table 9). Also consider computational complexity.

[BH-P8] See [BH-P1]. Evaluate performance impact of multi-parameter Intra prediction vs HEVC with enlarged CU and TU sizes (see Table 11). Also consider computational complexity.

[BH-P9] See [BH-P1]. Evaluate Joint performance impact of all tested tools on top of HEVC (see Table 6). Also consider computational complexity.

[BH-P10] Similar to [BH2], several coding tools (some of them were considered earlier during the initial development of HEVC) are investigated for NGVC based on HEVC [BH1]. These tools include large CTU and TU, adaptive loop filter, advanced temporal Motion Vector Prediction, cross component prediction, overlapped block Motion Compensation and adaptive multiple transform. The overall coding performance improvement resulting from these additional tools for classes A through F Video test sequences in terms of BD-rate are listed in Table 3. Class F sequences include synthetic (computer generated) video. Performance improvement of each tool for Random Access, All Intra and low-delay B is also listed in Table 4. Evaluate the performance improvement of these tools [BH1] over HM 16.4 and verify the results shown in Table 3.

[BH-P11] See [BH-P10]. Evaluate the performance improvement of each tool for All Intra, Random Access and Low-delay B described in Table 4. Test sequences for screen content coding (similar to class F in Table 3) can be accessed from http://pan.baidu.com/share/link?shareid=3128894651&uk=889443731.

[BH-P12] See [BH-P10] and [BH-P11]. Consider implementation complexity as another metric, as these additional tools invariably result in increased complexity. Evaluate this complexity for all classes (A through F) – see Table 3 and for All Intra, Random Access and low-delay B cases (see Table 4). See the papers below related to proposed requirements for Next Generation Video coding (NGVC).

1. ISO/IEC JTC1/SC29/WG11, "Proposed Revised Requirements for a Future Video coding Standard", MPEG doc.M36183, Warsaw, Poland, June. 2015.
2. M. Karczewicz and M. Budagavi, "Report of AHG on Coding Efficiency Improvements," VCEG-AZ01, Warsaw, Poland, June 2015.
3. J.-R. Ohm et al., "Report of AHG on Future Video Coding Standardization Challenges," MPEG Document M36782, Warsaw, Poland, June. 2015.

[BH-P13] In [BH1] for larger resolution video such as 4K and 8K, 64 × 64 INT DCT (integer approx.) is proposed to collaborate with the existing transforms (up to 32 × 32 INT DCT) and to further improve the coding efficiency. Madhukar and Sze developed unified forward and inverse transform architecture (2D-32 × 32 INT DCT) for HEVC (see IEEE ICIP 2012) resulting in simpler hardware compared to implementation separately i.e., forward and inverse. See also Chapter 6 HEVC transform and quantization by Budagavi, Fuldseth and Bjontegaard in [E202]. Implement similar uniform forward and inverse transform architecture for 2D-64 × 64 INT DCT and evaluate the resulting simpler hardware compared to implementation separately i.e., forward and inverse.

[BH-P14] Refer to Figures 6.7 and 6.8 of chapter 6 cited in [BH-P13]. These QM are based on visual sensitivity of the transform coefficients. Similar QM have been proposed/adopted in JPEG, MPEG-1, 2, 4, AVS china etc. Also see the book by K.R. Rao and P. Yip, "Discrete Cosine Transform", Academic press 1990, wherein the theory behind developing the QM is explained. See also related references at the end of chapter 6 in [E202]. The QM matrices (Figure 6.8) for (16 × 16) and (32 × 32) transform block sizes are obtained by replicating the QM for (8 × 8) transform block size. These extensions are based on reducing the memory needed to store them. Develop QM for (16 × 16) and (32 × 32) transform block sizes independently based on their visual sensitivities (perception).

[BH-P15] See [BH-P14]. Develop QM for (64 × 64) INTDCT reflecting the visual perception of these transform coefficients. Again refer to the book by K.R. Rao and P. Yip, "Discrete Cosine Transform", Academic press 1990.

[BH-P16] See [BH-P13] and [BH-P14]. In Figure 6.2 (page 149) of Chapter 6 [E202] (32 × 32) INTDCT, (4 × 4), (8 × 8), (16 × 16) INTDCTs are embedded. Develop (64 × 64) INTDCT wherein the smaller size INTDCTs are embedded. Is it orthogonal? What are the norms of the 64 basis vectors?

[BH-P17] See "Y. Sugito et al., "A study on addition of 64x64 transform to HM 3.0.," Joint Collaborative Team on Video Coding (JCT-VC) of ITU-T SG16 WP3 and ISO/IEC JTC1/SC29/WG11, JCTVC-F192, Torino, Italy, July, 2011" which is reference [15] in Alshin et al. [BH2]. Is the (64 × 64) INTDCT orthogonal? Embedding property?

[BH-P18] See [BH3]. Theoretical and implementation aspects of bi-directional optical flow which is part of JEM1 (Joint Exploration Model) See [2] in this paper are developed. Evaluate this MV refinement in terms of MC prediction compared to the ME process adopted in HEVC based on some standard test sequences. Sum of absolute MC prediction errors for various block sizes can be used as the comparison metric.

[BH-P19] In [BH4] enhanced multiple transform (EMT) is proposed and implemented on top of the reference software using various test sequences. See Table 4. This scheme is compared with HM-14.0 Main 10 for all intra (AI) and random access (RA) configurations. Verify these results and extend the simulation to low delay case. Use the latest HM software.

Post – HEVC Activity

Both MPEG and VCEG have established AHGs (ad hoc groups) for exploring next generation video coding.

Grois et al. (See item 7 under tutorials) have suggested as follows:

Focusing on perceptual models and perceptual quality, and perceptually optimized video compression provision: http://www.provision-itn.eu

PROVISION is a network of leading academic and industrial organizations in Europe including international researchers working on the problems with regard to the state-of-the-art video coding technologies.

The ultimate goal is to make noteworthy technical advances and further improvements to the existing state-of-the-art techniques of compression video material.

AV1 Codec

Alliance for Open Media (AOM) http://tinyurl.com/zgwdo59
Website: http://www.streamingmedia.com/Articles/Editorial/What-Is.../What-is-AV1-111497.aspx

AOM (charter members Amazon, Cisco, Google, Intel Corporation, Microsoft, Mozilla, and Netflix). In April 2016, ARM, AMD and NVIDIA joined the alliance to help ensure that the codec is hardware friendly and to facilitate and accelerate AV1 hardware support.

AOM has nearly finalized a codec called AV1 based on CISCO's Thor, Google's VP10 and Mozilla's DAALA with following goals:

- Interoperable and open
- Optimized for the web membership@aomedia.org
- Scalable to any modern device at any bandwidth
- Designed with a low computational footprint and optimized for hardware
- Capable of consistent, highest-quality, real-time video delivery
- Flexible for both commercial and non-commercial content, including user-generated content.

AV1 is to be shipped sometime between Dec. 2016 and March 2017 and is positioned to replace Google's VP9 and to compete with HEVC. It is designed to be royalty free.

In terms of makeup, the Alliance members enjoy leading positions in the following markets:

- Codec development – Cisco (Thor), Google (VPX), Mozilla (Daala)
- Desktop and mobile browsers – Google (Chrome), Mozilla (Firefox), Microsoft (Edge)
- Content – Amazon (Prime), Google (YouTube), Netflix
- Hardware co-processing - AMD (CPUs, graphics), ARM (SoCs, other chips), Intel (CPUs), NVIDIA (SoC, GPUs)
- Mobile – Google (Android), Microsoft (Windows Phone)
- OTT – Amazon (Amazon Fire TV), Google (Chromecast, Android TV)

The Alliance is targeting an improvement of 50 percent over VP9/HEVC with reasonable increases in encoding and playback complexity. One focus is UHD video, including higher bitrate, wider color gamut, and increased frame rates, with the group targeting the ability to play 4K 60fps in a browser on a reasonably fast computer. The base version of the codec will support 10-bit and 12-bit encoding, as well as the BT.2020 color space. Another focus is providing a codec for WebRTC (Real-Time Communications), an initiative supported by Alliance members Google and Mozilla, and similar applications including Microsoft's Skype.

P.S: This material on AV1 codec is collected from Streaming Media Magazine, "What is AV1", Jan Ozer, Posted on June 3, 2016.
For details on AOM, please access http://aomedia.org
The code base for AOM is here
https://aomedia.googlesource.com/

aomedia Git repositories - Git at Google aomedia.googlesource.com All-Projects Access inherited by all other projects. All-Users Individual user settings and preferences. aom Alliance for Open Media contributor-guide

The following link shows how to build and test AOM, and also provides test sequences.
http://aomedia.org/contributor-guide/

Alliance for Open Media – Contributor Guide
aomedia.org
Download Test Scripts. Test scripts (and this document) are available in the contributor-guide Git project: $ git clone https://aomedia.googlesource.com/contributor ...

AV1-P1 As AV1 is designed to be a royalty free codec, gather all details about this codec (website, streaming media etc.) and implement using standard test sequences. Compare its performance with HEVC using the standard criteria such as PSNR, BD bitrate, BD PSNR, SSIM, implementation complexity etc.

AV1-P2 See AV1-P1 Based on the performance comparison, explore any changes in the functionalities (transform/quantization/ME-MC, entropy coding, intra-inter and other modes) of the codec that can result in further improving AV1's compression performance with negligible increase in complexity.

AV1-P3 See AV1-P1 Can AV1 codec be applied to SCC? If so compare its performance with HEVC SCC extension.

AV1-P4 See AV1-P1 Explore how AV1 Codec can be modified for scalable video coding.

AV1-P5 See AV1-P1 Explore how AV1 Codec can be modified for 3D-Multiview coding.

AV1-P6 Performance comparison of SDCT with DCT and DDCT is investigated in [E392]. In AV1 INTDCT is used. Replace INTDCT with INTSDCT and implement the performance comparison based on various test sequences including 4K and 8K and different block sizes. See the Tables and Figures in [E392].

AV1-P7 See [E393]. Apply the FCDR to AV1 codec both as a post processing operation and as an in loop (embedded) operation. Implement the projects for AV1 similar to those described in P.5.268 and P.5.269.

Real Media HD (RMHD)

https://www.realnetworks.com/realmediaHD

Real Media HD, successor to its RMVB (Real Media Variable Bitrate) video codec.

Real Networks claims that RMHD codec has achieved a reduction of more than 30% bit rate compared to H.264/AVC, while achieving the same subjective quality.

Performance comparison of RMHD with H.264, H.265 and VP9 is also described.

RMHD-P1 While the RMHD encoder/decoder details are not known, develop comparison data similar to those shown in charts 1–7. Hopefully, RMHD software can be accessed.

SMPTE

SMPTE ST 2042-1:200x 28/09/2009
Page 9 of 131 pages
Foreword

SMPTE (the Society of Motion Picture and Television Engineers) is an internationally recognized standards developing organization. Headquartered and incorporated in the United States of America, SMPTE has members in over 80 countries on six continents. SMPTE's Engineering Documents, including Standards, Recommended Practices and Engineering Guidelines, are prepared by SMPTE's Technology Committees. Participation in these Committees is open to all with a bona fide interest in their work. SMPTE cooperates closely with other standards developing organizations, including ISO, IEC and ITU.

SMPTE Engineering Documents are drafted in accordance with the rules given in Part XIII of its Administrative Practices.

http://www.smpte.org E-mail: eng@smpte.org

About the Society of Motion Picture and Television Engineers® (SMPTE®) For the past 100 years, the people of the Society of Motion Pictures and Television Engineers (SMPTE, pronounced "simp-tee") have sorted out the details of many significant advances in entertainment technology, from the introduction of "talkies" and color television to HD and UHD (4K, 8K) TV. Since its founding in 1916, the Society has earned an Oscar® and multiple Emmy® Awards for its work in advancing moving-imagery education and engineering across the communications, technology, media, and entertainment industries. The Society has developed thousands of standards, recommended practices, and engineering guidelines, more than 800 of which are currently in force.

SMPTE's global membership today (Jan. 2017) includes 7,000 members, who are motion-imaging executives, engineers, creative and technology professionals, researchers, scientists, educators, and students. A partnership with the Hollywood Professional Association (HPA®) connects SMPTE and its membership with the professional community of businesses and individuals who provide the expertise, support, tools, and infrastructure for the creation and finishing of motion pictures, television programs, commercials, digital media, and other dynamic media content. Information on joining SMPTE is available at www.smpte.org/join.

Also SMPTE WEBCASTS (webinars).

VC-1

Video coding 1 is based on Windows Media Video (WMV9) developed by Microsoft. See [VC-1, 8] cited below for details.

References

VC-1, 1. H. Kalva and J.-B. Lee, "The VC-1 video coding standard", IEEE Multimedia, vol. 14, pp. 88–91, Oct.–Dec. 2007.

Vc-1, 2. J.-B. Lee and H. Kalva, "The VC-1 and H.264 video compression standards for broadband video services", Springer, 2008.

VC-1, 3. VC-1 Software : http://www.smpte.org/home

VC-1, 4. Microsoft website - VC-1 Technical Overview http://www.microsoft.com/windows/windowsmedia/howto/articles /vc1techoverview.aspx#VC1ComparedtoOtherCodecs

VC-1, 5. VC-1 Compressed video bitstream format and decoding process *(SMPTE 421M-2006)*, SMPTE Standard, 2006.

VC-1, 6. J.-B. Lee and H. Kalva, "An efficient algorithm for VC-1 to H.264 video transcoding in progressive compression", IEEE International Conference on Multimedia and Expo, pp. 53–56, July 2006.

VC-1, 7. S. Srinivasan and S. L. Regunathan, "An overview of VC-1" *Proc. SPIE*, vol. 5960, pp. 720–728, 2005.

VC-1, 8. S. Srinivasan et al., "Windows Media Video 9: overview and applications" Signal Processing: Image Communication, Vol. 19, pp. 851–875, Oct. 2004.

VC1-P1 See [E393]. Apply the FCDR to VC1 codec both as a post processing operation and as an in loop (embedded) operation. Implement the projects for VC1 similar to those described in P.5.268 and P.5.269.

WMV-9 (VC-1)

WMV9 of Microsoft

Windows Movie Maker, in Windows Vista or XP can be used, but with limited functionality. It can be used to convert a WMV file to AVI format, for example.

For more precise control of encoding parameters, you can use the Windows Media encoder tool, freely available at
http://www.microsoft.com/windows/windowsmedia/forpros/encoder/default.mspx

It is an encoding-only tool.

Source code for such tools is not available.

You can also use commercial video editing tools, such as the ones listed at

http://www.microsoft.com/windows/windowsmedia/forpros/service_provider/software/default.aspx#encoding

Source Code for VC-1 (WMV9)

http://store.smpte.org/VC-1-Test-Material-p/vc-1.htm

Please note: We must receive your completed VC-1 License Agreement Form before we can ship your VC-1 CD-ROM. You can download this form from the SMPTE Website at:
http://www.smpte.org/standards/.
After you've completed the form please fax back to SMPTE 914-761-3115 or email a scanned copy to aseminara@smpte.org
Our Price: $450.00

1. The below link had a lot of information on audio and video codecs. http://wiki.multimedia.cx/index.php?title=Main_Page
2. The link below has a VC-1 reference decoder. It is not the reference decoder but a free one!
   http://ffmpeg.mplayerhq.hu/download.html
3. Also, Google search provided the reference code for VC 1 – version 6 which might be outdated now and may not be legal to use (I am not sure on this)PersonName. http://www.richardgoodwin.com/VC1_reference_decoder_release6.zip

VC-2 Standard

The VC-2 standard specifies the compressed stream syntax and reference decoder operations for a video compression system. VC-2 is an intra frame video compression system aimed at professional applications that provides efficient coding at many resolutions including various flavors of CIF, SDTV and HDTV. VC-2 utilizes wavelet transforms that decompose the video signal into frequency bands. The codec is designed to be simple and flexible, yet be able to operate across a wide range of resolutions and application domains.

The system provides the following capabilities:

- **Multi-resolution transforms**. Data is encoded using the wavelet transform, and packed into the bitstream subband by subband. High compression ratios result in a gradual loss of resolution. Lower resolution output pictures can be obtained by extracting only the lower resolution data.
- **Frame and field coding**. Both frames and fields can be individually coded.
- **CBR and VBR operation**. VC-2 permits both constant bit rate and variable bit rate operations. For low delay pictures, the bit rate will

be constant for each area (VC-2 slice) in a picture to ensure constant latency.

- **Variable bit depths**. 8, 10, 12 and 16 bit formats and beyond are supported.
- **Multiple color difference sampling formats**. 444, 422 and 420 video are all supported.
- **Lossless and RGB coding**. A common toolset is used for both lossy and lossless coding. RGB coding is supported either via the YCoCg integer color transform for maximum compression efficiency, or by directly compressing RGB signals.
- **Wavelet filters**. A range of wavelet filters can be used to trade off performance against complexity. The Daubechies (9, 7) filter is supported for compatibility with JPEG2000. A fidelity filter is provided for improved resolution scalability.
- **Simple stream navigation**. The encoded stream forms a doubly-linked list with each picture header indicating an offset to the previous and next picture, to support field-accurate high-speed navigation with no parsing or decoding required.
- **Multiple Profiles**. VC-2 provides multiple profiles to address the specific requirements of particular applications. Different profiles include or omit particular coding tools in order to best match the requirements of their intended applications. The Main profile provides maximum compression efficiency, variable bit rate coding and lossless coding using the core syntax. The Simple profile provides a less complex codec, but with lower compression efficiency, by using simple variable length codes for entropy coding rather than the arithmetic coding used by the Main profile. The Low Delay profile uses a modified syntax for applications requiring very low, fixed, latency. This can be as low as a few lines of input or output video. The Low Delay profile is suitable for light compression for the re-use of low bandwidth infrastructure, for example carrying HDTV over SD-SDI links. The High Quality profile similarly provides light compression with low latency and also supports variable bit rate and lossless coding.

Scope

This standard defines the VC-2 video compression system through the stream syntax, entropy coding, coefficient unpacking process and picture decoding process. The decoder operations are defined by means of a mixture of pseudocode and mathematical operations.

VC2-P1 Compare the performance of VC2 lossless coding with other lossless coding techniques described in JPEG/JPEG-LS, JPEG-XR, JPEG-XT, JPEG2000, HEVC, H.264, MPEG-2 and DIRAC-PRO based on standard comparison metrics.

VC2-P2 Performance comparison of SDCT with DCT and DDCT is investigated in [E392]. In VC2 replace wavelets with INTSDCT and implement the performance comparison based on various test sequences including 4K and 8K and different block sizes. See the Tables and Figures in [E392].

ATSC (Advanced Television Systems Committee)

The Advanced Television Systems Committee, Inc., is an international, non-profit organization developing voluntary standards for digital television. The ATSC member organizations represent the broadcast, broadcast equipment, motion picture, consumer electronics, computer, cable, satellite, and semiconductor industries.

Specifically, ATSC is working to coordinate television standards among different communications media focusing on digital television, interactive systems, and broadband multimedia communications. ATSC is also developing digital television implementation strategies and presenting educational seminars on the ATSC standards.

ATSC was formed in 1982 by the member organizations of the Joint Committee on InterSociety Coordination (JCIC): the Electronic Industries Association (EIA), the Institute of Electrical and Electronic Engineers (IEEE), the National Association of Broadcasters (NAB), the National Cable Telecommunications Association (NCTA), and the Society of Motion Picture and Television Engineers (SMPTE). ATSC members represent the broadcast, broadcast equipment, motion picture, consumer electronics, computer, cable, satellite, and semiconductor industries.

International adopters of the ATSC standard include Canada, Dominican Republic, El Salvador, Guatemala, Honduras, Mexico, and South Korea.

ATSC 3.0 CANDIDATE STANDARDS (Dec. 2016)

Next generation broadcast TV

The A/341 Video standard specifies how to encode video in the ATSC 3.0 system. It uses HEVC (H.265) video compression, the latest MPEG video coding standard, and provides support for 4K Ultra HDTV as well as future capabilities for transmission of information to enable wider color gamut, higher frame rates and high dynamic range. This is a candidate standard.

Legacy Codec

LC1. J. Chen et al., "Efficient Video Coding Using Legacy Algorithmic Approaches," IEEE Trans. on Multimedia, vol. 14, pp. 111–120, Feb. 2012.

LC-P1. See [LC1]. The authors have developed a video codec based on the traditional transform predictive coding aided by motion compensation followed by the QM-coder. While its performance is similar to that of H.264/AVC, the main advantage is its royalty few feature. Implement this legacy codec using various test sequences including ultra HD TV resolution.

DSC by VESA

Display stream compression (DSC) standard developed by Video Electronics Standards Association (VESA) addresses low-cost (light weight compression) visually lossless video codec that meets the requirements of display links. The papers by Walls and MacInnis [DSC1, DSC2] provide the background and an overview of DSC problem and its history. Both the algorithmic compression and performance impact of some tools and test results are presented. As detailed description of DSC is beyond the scope of this book, relevant references and pertinent web sites [DSC3–DSC5] are listed. Also VESA issued a call for technology in order to standardize a significantly more complex codec called advanced DSC (ADSC) [DSC5] that is visually lossless at a lower bit rate than DSC. Peng et al. [OP11] in the overview paper on SCC present not only the details on DSC but also discuss on going work and future outlook on SCC and DSC.

DSC1 F.G. Walls and A.S. MacInnis, "VESA display stream compression for television and cinema applications', IEEE JETCAS, vol. 6, issue 4, pp. 360–370, Dec. 2016.

DSC2 MF. Walls and A. S. MacInnis, "VESA display stream compression: An overview," in *SID Symp. Dig. Tech. Papers*, Jun. 2014, vol. 4. no. 1, pp. 360–363.

DSC3 Video Electronics Standards Association. (2016). *VESA Display Stream Compression (DSC) Standard v1.2.* [Online]. Available: http://vesa.org

DSC4 (Jan. 2013). *VESA Finalizes Requirements for Display Stream Com pression Standard*. [Online]. Available: http://www.vesa.org/news/vesa-finalizes-requirements-for-display-stream-compression-standard/_press.html

DSC5 (Jan. 2015). *VESA Issues Call for Technology: Advanced Display Stream Compression*. [Online]. Available: http://www.vesa.org/news/vesa-issues-call-for-technology-advanced-display-stream-compression/

DSC6 VESA Updates Display Stream Compression Standard to Support New Applications and Richer Display Content, Jan. 2016 [online] Available: http://www.vesa.org/featured-articles/vesa-updates-displaystream-compression-standard-to-support-new-applications-and-richerdisplay-content

DSC-P1 Go through the review papers on DSC [DSC1, DSC2] and confirm the performance results shown in Tables II–V using the corresponding test images. This project is fairly complex and may require group effort.

DSC-P2 Access the web site on call for technology for ADSC [DSC5] and contribute some proposals that can meet the requirements of ADSC. Follow this up in terms of comparing your proposals with proposals submitted by other groups.

DSC-P3 See [DSC6] and access the web site. Explore how the DSC standard can support new applications and richer display content.

PSNRAVG

PSNRAVG is a weighted average of luminance (PSNRY) and chrominance (PSNRU and PSNRV) PSNR components. All involved test sequences are in 4:2:0 color format, for which PSNRAVG is computed as (de facto standard).

PSNRAVG = (6 × PSNRY + PSNRU + PSNRV)/8.

Since PSNRAVG also takes the impact of the chrominance components into account, it is supposed to provide more reliable results than the conventional PSNRY metric in the cases when the luminance and chrominance components have dissimilar RD behaviors. See reference below:

B. Li, G. J. Sullivan, and J. Xu, RDO with Weighted Distortion in HEVC, document JCTVC-G401, ITU-T/ISO/IEC Joint Collaborative Team on Video Coding (JCT-VC), Geneva, Switzerland, Nov. 2011.

Index

About the Authors

K. R. Rao received the Ph.D. degree in electrical engineering from The University of New Mexico, Albuquerque in 1966. He is now working as a professor of electrical engineering in the University of Texas at Arlington, (UTA) Texas. He has published (coauthored) 19 books, some of which have been translated into Chinese, Japanese, Korean, Spanish and Russian. Also as e-books and paper back (Asian) editions. He has supervised 113 Masters and 31 doctoral students. He has published extensively and conducted tutorials/ workshops worldwide. He has been a visiting professor in National university of Singapore and electronics and telecommunications research institute (ETRI), Taejon, Korea. He has been a keynote speaker in many national and international conferences. He has been a consultant to academia, industry and research institutes. He has been an external examiner for several M.S. and Ph.D. students worldwide. He has been a reviewer of research proposals from Brazil, China, India, Korea, Singapore, Taiwan, Thailand and US. He was invited to review applications for recruitment and/or promotion of faculty in various Universities (US and abroad). He is an IEEE Fellow. He has been a member of the academy of distinguished scholars, UTA.

Jae-Jeong Hwang is currently the director of Innovation Center for Engineering Education at Kunsan National University, Korea. He is also professor of School of IT Information and Control Engineering at the KNU, since 1987 and adjunct professor of RMIT University, Melbourne, Australia, since 2008.

He served as the dean of Engineering College, the dean of graduate school of industry, the director of Information & Computing Center, Engineering Research Center, BK21 Industry-Academy Cooperated Education Project, and Small & Medium Enterprises Center.

His research interests are digital image/video coding & processing, information theory, object segmentation and tracking, and IT convergence and applications. He is the coauthor of *Techniques and standards for image, video and audio coding* (Prentice Hall, 1996), *Digital Image/Video Engineering* (Ajin, 1999), *Digital Television Master* (InterVision, 2007), *Information and Coding Tech.* (Dooyangsa, 2008), and *Fast Fourier transform – Algorithms and applications* (Springer, 2010), *Video coding standards: AVS China, H.264/MPEG-4 PART 10, HEVC, VP6, DIRAC and VC-1* (Springer, 2014). He has also written some chapters in the books on digital video image quality and perceptual coding and digital consumer electronics handbook.

Do Nyeon Kim received the Ph.D. degree in electrical and electronic engineering from Yonsei University in Seoul, South Korea in 2004. From 1989 to 2003, he was with the Electronics and Telecommunications Research Institute (ETRI), South Korea, where he was a senior researcher. From 2005 to 2010, he was with the University of Texas at Arlington where he was a visiting scholar of electrical engineering. Since 2010, he has been with Barun Technologies, Corp., South Korea, where he is currently a senior engineer. He has published "*Fast Fourier Transform – Algorithms and Applications*" (with K. R. Rao and J. J. Hwang, Springer, 2010). This has been translated into Chinese and Korean. Also *"Video coding standards: AVS China, H.264/MPEG-4 PART 10, HEVC, VP6, DIRAC and VC-1",* (with K. R. Rao and J. J. Hwang, Springer, 2014). This has been translated into Chinese.